T0140145

# Studies in Computational Intelligence

Volume 653

**Series editor**

Janusz Kacprzyk, Polish Academy of Sciences, Warsaw, Poland
e-mail: kacprzyk@ibspan.waw.pl

*About this Series*

The series "Studies in Computational Intelligence" (SCI) publishes new developments and advances in the various areas of computational intelligence—quickly and with a high quality. The intent is to cover the theory, applications, and design methods of computational intelligence, as embedded in the fields of engineering, computer science, physics and life sciences, as well as the methodologies behind them. The series contains monographs, lecture notes and edited volumes in computational intelligence spanning the areas of neural networks, connectionist systems, genetic algorithms, evolutionary computation, artificial intelligence, cellular automata, self-organizing systems, soft computing, fuzzy systems, and hybrid intelligent systems. Of particular value to both the contributors and the readership are the short publication timeframe and the worldwide distribution, which enable both wide and rapid dissemination of research output.

More information about this series at http://www.springer.com/series/7092

Roger Lee
Editor

# Software Engineering, Artificial Intelligence, Networking and Parallel/Distributed Computing

 Springer

*Editor*
Roger Lee
Software Engineering and Information
Central Michigan University
Mount Pleasant
USA

ISSN 1860-949X                    ISSN 1860-9503   (electronic)
Studies in Computational Intelligence
ISBN 978-3-319-81608-1           ISBN 978-3-319-33810-1   (eBook)
DOI 10.1007/978-3-319-33810-1

Printed on acid-free paper

This Springer imprint is published by Springer Nature
The registered company is Springer International Publishing AG Switzerland

# Preface

The purpose of the 17th IEEE/ACIS International Conference on Software Engineering, Artificial Intelligence, Networking and Parallel/Distributed Computing (SNPD 2016) held during May 30–June 1, 2016 in Shanghai, China, is aimed at bringing together researchers and scientists, businessmen and entrepreneurs, teachers and students to discuss the numerous fields of computer science, and to share ideas and information in a meaningful way. This publication captures 14 of the conference's most promising papers, and we impatiently await the important contributions that we know these authors will bring to the field.

In Chap. "Dealing with Missing Values in Software Project Datasets: A Systematic Mapping Study", Ali Idri, Ibtissam Abnane, and Alain Abran carried out a systematic mapping study to summarize the existing techniques dealing with MV in SE datasets and to classify the selected studies according to six classification criteria. The study shows an increasing interest in machine learning (ML) techniques.

In Chap. "Testing and Code Review Based Effort-Aware Bug Prediction Model", K. Muthukumaran, N.L. Bhanu Murthy, G. Karthik Reddy, and Prateek propose an effort-aware bug prediction model that will consider the effort required to perform the specific quality assurance activity and empirically prove the supremacy of these models over the existing effort-aware models.

In Chap. "Improving the Efficiency of Schedulability Tests for Fixed Priority Preemptive Systems", Guangliang Yu, Mengfei Yang, Hejie Sun, and Hong Jiang discuss the way of how to find better initial values for sufficient schedulability tests. Extensive evaluations show that the schedulability tests found by our method need less runtime than state-of-the-art approaches.

In Chap. "A GPU Accelerated Finite Differences Method of the Bioheat Transfer Equation for Ultrasound Thermal Ablation", Georgios Kalantzis, Warner Miller, Wolfgang Tichy, and Suzanne LeBlang propose a method for high intensity focused ultrasound (FUS) therapy with three GPUs.

In Chap. "Justifying the Transition from Trustworthiness to Resiliency via Generation of Safety Cases", Chung-Ling Lin, Wuwei Shen, and Steven Drager propose a novel approach to show how an argument structure can be automatically built via safety case patterns and metamodels underlying a development process.

In Chap. "Select and Test (ST) Algorithm for Medical Diagnostic Reasoning", D.A. Irosh P. Fernando and Frans A. Henskens present an enhanced version of the ST Algorithm. The focus of this paper is on the algorithm, which is intended to give a theoretical proof that medical expert systems are achievable.

In Chap. "Opportunities Ahead the Future Mobile Learning", Regin Joy Conejar, Haeng-Kon Kim, and Roger Y. Lee discuss the opportunities ahead the future mobile learning. The paper reflects on the innovation and the complexities that are currently emerging in education as a result of these technological advancements.

In Chap. "A Design of Context-Aware Framework for Conditional Preferences of Group of Users", Reza Khoshkangini, Maria Silvia Pini, and Francesca Rossi propose a context-aware framework that provides service(s) according to the current context of entities and the current users' preferences, which are naturally conditional.

In Chap. "Design of Decentralized Inter-Cell Interference Coordination Scheme in LTE Downlink System", Yen-Wen Chen and KangHao Lo propose a decentralized ICIC algorithms, named relative throughput based resource block coordination (RTRBC) and RTRBC with residual RB (RTRBC_r), to deal with the coordination among eNBs within the existing LTE framework.

In Chap. "Toward Flow-Based Ontology", Sabah Al-Fedaghi introduces an exploratory study of ontology multi-relationships with the aim of adding a more systematic foundation that visualizes structure and incorporates procedures in the form of input-process-output.

In Chap. "Mobile Component Integration Agent (MCIA) for Social Business Application", Yvette E. Gelogo, Haeng-Kon Kim, and Roger Y. Lee propose a mobile component integration agent (MCIA) for social business application as a software development methodology to simply integrate the different technology building blocks into one web-based solution.

In Chap. "Covariance Estimation for Vertically Partitioned Data in a Distributed Environment", Aruna Govada and S.K. Sahay propose a communication-efficient algorithm to estimate the covariance matrix in a distributed manner. The results show that it is exactly same as centralized method with good speed-up in terms of computation.

In Chap. "Detection of Dengue Epidemic in Dhaka, Bangladesh by a Neuro Fuzzy Approach", Md. Arifuzzaman, Md. Faqrul Islam Shaon, Md. Jahidul Islam, and Rashedur M. Rahman research aims to develop an efficient system which will identify the probability of dengue occurrence in Dhaka, Bangladesh based on a neural network system and fuzzy inference algorithm.

In Chap. "MIMO Antenna Design for Future 5G Wireless Communication Systems", M. Aziz ul Haq, M. Arif Khan, and Md. Rafiqul Islam presents a novel multiple input multiple output (MIMO) antenna design for future 5G wireless communication systems. An ultra-wideband antenna array is designed on a printed circuit board which is suitable for radio frequency circuits.

It is our sincere hope that this volume provides stimulation and inspiration, and that it will be used as a foundation for works to come.

May 2016

Yihai Chen
Shanghai University, China

# Contents

# Contributors

**Ibtissam Abnane** Software Project Management Research Team, ENSIAS, Mohamed V University, Rabat, Morocco

**Alain Abran** Department of Software Engineering, Ecole de Technologie Supérieure, Montréal, Canada

**Sabah Al-Fedaghi** Computer Engineering Department, Kuwait University, Safat, Kuwait

**M. Arif Khan** School of Computing and Mathematics, Charles Sturt University, Wagga, NSW, Australia

**Md. Arifuzzaman** Department of Electrical and Computer Engineering, North South University, Bashundhara, Dhaka, Bangladesh

**M. Aziz ul Haq** School of Computing and Mathematics, Charles Sturt University, Wagga, NSW, Australia

**N.L. Bhanu Murthy** BITS Pilani Hyderabad Campus, Hyderabad, India

**Yen-Wen Chen** Department of Communication Engineering, National Central University, Tao-Yuan, Jhong-Li, Taiwan

**Regin Joy Conejar** School of Information Technology, Catholic University of Daegu, Daegu, Korea

**Steven Drager** Air Force Research Laboratory, Rome, NY, USA

**Md. Faqrul Islam Shaon** Department of Electrical and Computer Engineering, North South University, Bashundhara, Dhaka, Bangladesh

**D.A. Irosh P. Fernando** Distributed Computing Research Group, School of Electrical Engineering and Computer Science, School of Medicine and Public Health, University of Newcastle, Callaghan, NSW, Australia

**Yvette E. Gelogo** School of Computing and Mathematics, Charles Sturt University, Daegu, Korea

**Aruna Govada** BITS-Pilani, Zuarinagar, Goa, India

**Frans A. Henskens** Distributed Computing Research Group, Health Behaviour Research Group, School of Electrical Engineering and Computer Science, University of Newcastle, Callaghan, NSW, Australia

**Ali Idri** Software Project Management Research Team, ENSIAS, Mohamed V University, Rabat, Morocco

**Md Rafiqul Islam** School of Computing and Mathematics, Charles Sturt University, Wagga, NSW, Australia

**Md. Jahidul Islam** Department of Electrical and Computer Engineering, North South University, Bashundhara, Dhaka, Bangladesh

**Hong Jiang** Beijing Institute of Control Engineering, Beijing, China

**Georgios Kalantzis** Department of Physics, Florida Atlantic University, Boca Raton, FL, USA

**G. Karthik Reddy** Stony Brook University, New York, USA

**Reza Khoshkangini** Department of Mathematics, University of Padova, Padua, Italy

**Haeng-Kon Kim** School of Information Technology, Catholic University of Daegu, Daegu, Korea

**Suzanne LeBlang** University MRI & Diagnostic Imaging Centers, Boca Raton, FL, USA

**Roger Y. Lee** Department of Computer Science, Central Michigan University, Michigan, USA

**Chung-Ling Lin** Department of Computer Science, Western Michigan University, Kalamazoo, MI, USA

**Kang-Hao Lo** Department of Communication Engineering, National Central University, Tao-Yuan, Jhong-Li, Taiwan

**Warner Miller** Department of Physics, Florida Atlantic University, Boca Raton, FL, USA

**K. Muthukumaran** BITS Pilani Hyderabad Campus, Hyderabad, India

**Maria Silvia Pini** Department of Information Engineering, University of Padova, Padua, Italy

**Rashedur M. Rahman** Department of Electrical and Computer Engineering, North South University, Bashundhara, Dhaka, Bangladesh

**Francesca Rossi** IBM T. J. Watson Research Center, USA (on Leave from University of Padova), Yorktown Heights, NY, USA

**S.K. Sahay** BITS-Pilani, Zuarinagar, Goa, India

**Wuwei Shen** Department of Computer Science, Western Michigan University, Kalamazoo, MI, USA

**Hejie Sun** Beijing Institute of Control Engineering, Beijing, China

**Prateek Talishetti** BITS Pilani Hyderabad Campus, Hyderabad, India

**Wolfgang Tichy** Department of Physics, Florida Atlantic University, Boca Raton, FL, USA

**Mengfei Yang** China Academy of Space Technology, Beijing, China

**Guangliang Yu** Beijing Institute of Control Engineering, Beijing, China

# Dealing with Missing Values in Software Project Datasets: A Systematic Mapping Study

Ali Idri, Ibtissam Abnane and Alain Abran

**Abstract** Missing Values (MV) present a serious problem facing research in software engineering (SE) which is mainly based on statistical and/or data mining analysis of SE data. Therefore, various techniques have been developed to deal adequately with MV. In this paper, a systematic mapping study was carried out to summarize the existing techniques dealing with MV in SE datasets and to classify the selected studies according to six classification criteria: research type, research approach, MV technique, MV type, data types and MV objective. Publication channels and trends were also identified. As results, 35 papers concerning MV treatments of SE data were selected. This study shows an increasing interest in machine learning (ML) techniques especially the K-nearest neighbor algorithm (KNN) to deal with MV in SE datasets and found that most of the MV techniques are used to serve software development effort estimation techniques.

**Keywords** Systematic mapping study · Software engineering · Missing values

## 1 Introduction

Performing empirical studies in software engineering has increased recently with the availability of larger software project datasets, such as those of the International Software Benchmarking Standards Group—ISBSG and PROMISE [12, 24].

A. Idri (✉) · I. Abnane
Software Project Management Research Team,
ENSIAS, Mohamed V University, Rabat, Morocco
e-mail: idri@ensias.ma

I. Abnane
e-mail: Ibtissam.abnane19@gmail.com

A. Abran
Department of Software Engineering, Ecole de Technologie Supérieure,
Montréal H3C IK3L, Canada
e-mail: Alain.abran@etsmtl.ca

© Springer International Publishing Switzerland 2016
R. Lee (ed.), *Software Engineering, Artificial Intelligence, Networking and Parallel/Distributed Computing*, Studies in Computational Intelligence 653, DOI 10.1007/978-3-319-33810-1_1

1

Indeed, up until the early 2000s, most of the empirical research in SE field have been carried out with small samples. Nowadays, larger historical software project datasets are often used with a higher number of MV for a significant number of variables, therefore making their use rather challenging for research aims [4, 22]. Thereby, a proper handling of MV is necessary when analyses are performed in a domain, such as empirical SE, where accuracy and precision are key factors. Various approaches have been investigated to handle MV in SE datasets, such as (1) deletion techniques which delete the data with missing values from the datasets [2, 31]; (2) toleration techniques which perform analysis directly on incomplete data sets [28]; and (3) imputation techniques which firstly fill the incomplete features and then do analysis on complete data sets [13, 14, 22]. As the SE research area matures and the number of papers related to the issue increases, it is important to systematically identify, classify and analyze the state of art and provides an overview of the trend in the field of MV techniques in SE data sets. This paper presents a systematic mapping of MV techniques in SE datasets. To the best of our knowledge, this paper is the first systematic mapping study in the area of MV techniques in SE datasets. The main contributions of this paper are: (1) a classification scheme categorizing articles in the field of MV techniques in SE datasets; (2) a systematic mapping study of MV techniques in SE, structuring related research work over the past decade by analyzing 35 selected papers; (3) an analysis of the demographic trends in the area of MV techniques in SE datasets; and (4) a repository of the papers collected and analyzed through this systematic study. The results summarize the existing MV techniques in SE datasets, the types of MV treated, the type of data analyzed, and the objective investigated behind the use of MV techniques. The research types and approaches that exist in literature were also identified. The results were analyzed and synthesized to provide an updated and summarized view and a set of recommendations for researchers and practitioners. The structure of this paper is as follows: Sect. 2 presents the research method used in this study. Section 3 reports the results obtained from the systematic mapping study. Section 4 discusses the main findings. Section 5 outlines threats to validity. The conclusion is presented in Sect. 6.

## 2  Research Methodology

### 2.1  Mapping Questions

The aim of this study is to identify, analyze, and synthesize the work published during the past decade (2000–2013) in the field of MV techniques in SE datasets. Towards this aim, eight research questions as well as the rationale motivating their importance were raised as presented in Table 1 [23].

**Table 1** Mapping Questions

| ID | Research question | Motivation |
|----|------------------|------------|
| MQ1 | Which publication channels are the main targets for MV techniques in SE datasets research? | To identify where MV research can be found, in addition to the targets for the publication of future studies |
| MQ2 | How has the frequency of approaches related to MV techniques in SE datasets changed over time? | To identify the publication trends of MV techniques in SE datasets research over time |
| MQ3 | In which research types are MV techniques in SE datasets papers, classified? | To explore the different types of research reported in MV techniques in SE datasets literature |
| MQ4 | What are the research approaches of the selected papers? | Discover the research approaches most investigated when dealing with in MV for SE data sets |
| MQ5 | What MV techniques are investigated the most? | To gain knowledge about the type of MV techniques that have been the most popular to identify trends and possible shortcomings or opportunities for MV's model focus |
| MQ6 | What type(s) of data are treated by those techniques? | To identify and present types of data treated in the studies |
| MQ7 | What type of MV is investigated the most? | To identify and classify MV types investigated in the researches and identify trends and possible shortcomings for MV's type focus |
| MQ8 | What are the objectives of investigating MV techniques in SE datasets? | To gain knowledge about motivations behind investigating MV techniques in SE datasets |

## 2.2 Search Strategy

Primary studies were identified by consulting the six online libraries shown in Fig. 1. In order to construct the search string used to perform the search in the six digital libraries, we derived major terms from the mapping questions of Table 1 and checked for their synonyms and alternative spellings. The search query was formulated as follow:

*(Miss\* OR incomplet\*) **AND** (Value OR attribute OR data\* OR input OR variable OR feature) **AND** (experiment\* OR metric OR measur\* OR assess\* OR evaluat\* OR predict\*) **AND** (software OR application OR program OR system) **AND** (Engineering OR maintenance OR science OR develop\* OR test\* OR construct\* OR design\* OR project OR effort OR cost OR requirement OR quality OR process) **AND** (imput\* OR deal\*OR handl\*).*

Note that the terms related to software engineering mainly come from the *Guide to the Software Engineering Body of Knowledge (SWEBOK)* [1]. The search strategy was conducted in two iterations: One author firstly searched the six electronic databases by applying the search string to the title, abstract and keywords to

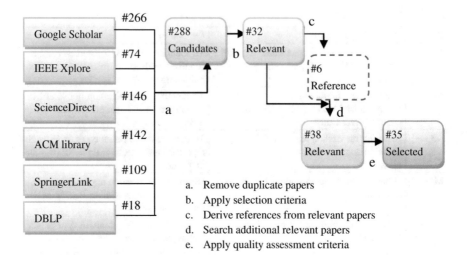

a.  Remove duplicate papers
b.  Apply selection criteria
c.  Derive references from relevant papers
d.  Search additional relevant papers
e.  Apply quality assessment criteria

**Fig. 1** Search and selection process

gather the candidate papers. Each paper was retrieved and the information about it was filed in the excel file that comprises the title of the study, its authors, source, abstract, keywords. Two authors considered each paper title, abstract and some-times the full text, and then commented on whether the paper should be "included", "excluded" or "uncertain" according to the inclusion and exclusion criteria of Sect. 2.3. If both researchers categorized one paper as "included", the paper was considered to be relevant; if both researchers categorized one paper as "Excluded", the paper was excluded; otherwise, the paper was labeled "uncertain", which means that the researchers disagreed on its relevance. Those papers were discussed by both researchers until an agreement was reached for each one. Then, we scanned the reference lists of relevant papers to find extra relevant papers and add them to the pool so that the risk of missing important publications can be reduced.

## 2.3  Study Selection

As can be seen in Fig. 1, the search in the six electronic databases resulted in 288 candidate papers. To identify selected papers, further screening was performed by applying two selection phases. Phase 1: Inclusion and exclusion criteria were used to identify the relevant papers. Phase 2: Quality assessment criteria of Sect. 2.4 were applied to the relevant papers of phase 1 so as to select the papers with acceptable quality, which were eventually used in data extraction. We defined the following inclusion and exclusion criteria, and then we carried out the study selection by reading the titles, abstracts, or full text of the papers.

- Inclusion criteria: (1) Use of a technique to deal with MV in SE datasets; (2) Comparison of two or more MV techniques using SE datasets; (3) Use of techniques for predicting and/or evaluating any software project attribute such as effort or quality which integrate one or more MV techniques; and/or (4) Where several papers reported the same study, only the most recent was included.
- Exclusion criteria: (1) Studies dealing with MV but are not in the SE domain; (2) Papers treating MV in contexts other than experimentation in SE; (3) Studies not available in full text; and/or (4) Studies not in English language.

By applying the inclusion and exclusion criteria in selection phase 1, we identified 32 relevant papers. Then, by scanning the references of these 32 relevant papers, 6 extra relevant papers were added in the initial search. Therefore, we identified in total 38 relevant papers. In phase 2, the quality assessment criteria were applied to the 38 relevant papers and 35 papers were finally selected which were then used in data extraction. The details of the quality assessment phase are described in Sect. 2.4.

## 2.4 Quality Assessment

The QA questionnaire contained six questions. Similar criteria were used in [11] and [23]:

1. Are the aims of the research clearly defined? Yes (+1) if the authors gives clear motivations of the study, Partially (+0.5) if the authors gives unclear motivations of the study, and No (+0) if the authors gives no motivation of the study.
2. Are the MV techniques well defined? Yes (+1) when the paper presents an overview and an explanation of the measures related, Partially (+0.5) when an overview is presented without explaining details and measures, and No (+0) when the paper uses methods already explained in previous works.
3. Is the experiment applied on large project data sets? We based the evaluation of the size of the datasets on the number of projects used in the study N. (Yes (+1) if $1000 \leq N$, Partially (+0.5) if $100 \leq N < 1000$ and No (+0) if $N < 100$).
4. Are MV techniques compared with others? Yes (+1), No (+0).
5. Are the findings of study clearly stated and supported by reporting results? Yes (+1) when results are explained and illustrated by examples, tables, and statistics, Partially (+0.5) when the findings are briefly presented, with insufficient illustrations, and No (+0) when results are presented without any illustrations.
6. Has the study been published in a recognized and stable journal or conference proceedings or workshop or in a symposium? This question was rated by considering the computer science conference rankings (CORE) (A*, A, B and C conferences), and the 2012 Journal Citation Reports (JCR) list. For conferences, workshops and symposia: (+1.5) if it is ranked CORE A, (+1) if it is ranked CORE B, (+0.5) if it is ranked CORE C, (+0) If it is not in CORE ranking. For journals: (+2) if it is ranked Q1, (+1.5) if it is ranked Q2, (+1) if it is ranked Q3 or Q4 (+0) If it has no JCR ranking. For others (+0).

**Table 2** Quality levels of relevant studies

| Quality level | # of studies | Percent (%) |
|---|---|---|
| Very high (6 ≤ score ≤ 7) | 8 | 21 |
| High (4.5 ≤ score < 6) | 19 | 50 |
| Medium (3.5 ≤ score < 4.5) | 8 | 21 |
| Low (2 ≤ score < 3.5) | 3 | 8 |
| Very low (0 ≤ score < 2) | 0 | 0 |
| Total | 38 | 100 |

Two researchers of this review performed the QA of the relevant studies individually. To ensure the quality of the findings of this study, we only considered the relevant studies with acceptable quality, i.e., with quality score equal or greater than 3.5 (50 % of perfect score), for the subsequent data extraction and data synthesis. Accordingly, we further dropped in Phase 2 any relevant papers with quality score less than 3.5. Since in the selection strategy we have required the quality score of studies must exceed 3.5, we believe that the selected studies are of high quality. Indeed, as shown in Table 2, about 71 % of the selected studies are of a high or very high quality level. Due to the limit number of pages, the details of the complete list of selected studies as well as their quality scores may be sent upon request to researchers for further investigations.

## 2.5   Data Extraction

The 35 selected studies were used to collect data that would provide the answers to the MQs of Table 1. The publication source and channel of the papers selected respond to MQ1, while the publication year responds to MQ2. Regarding MQ3, a research type can be classified as in [11, 23]: Evaluation Research, Solution Proposal or Other (e.g. Opinion papers). With regard to MQ4, the research approach can be [11, 23]: Case study, Survey, Experiment, History-based evaluation, Theory or Other. As regards to MQ5, a missing values treatment techniques can be classified as: missing data toleration [28], missing data deletion [31], or missing data imputation [13], among which, two large categories can be distinguished: (1) Non-Machine Learning imputation techniques(non-ML) such as Mean Imputation (MEI), Expectation Maximization Imputation (EMI), Regression Imputation (RI), Multiple Imputation (MI) or other (e.g. Event Covering Imputation (ECI), Full Information Maximum Likelihood Imputation (FIMLI); and (2) Machine Learning techniques (ML) such as Decision Trees Imputation(DTI), k-Nearest Neighbors Imputation (KNNI), or others (e.g., Most Common Attribute Imputation, Association Rules Imputation).With regards to MQ6, the type of data can be classified as Numerical or Categorical. With regard to MQ7, The mechanism causing the incompleteness can affect which imputation method could be applied. Generally, these mechanisms fall into three classes [15]: Missing Completely At Random (MCAR), Missing At Random (MAR), and Missing Not At Random (MNAR). The response to MQ8 can either be: (1) Software Effort Prediction (SEP): MV techniques are used with software development effort prediction techniques to deal with incomplete datasets, (2) No Software Effort Prediction (NSEP): MV techniques are used to evaluate and/or predict any software project attribute other than effort, and (3) MV Performance Evaluation (MVPE): MV techniques are using SE datasets to evaluate their performances without predicting or evaluating any software project attribute.

# 3 Results and Discussion

## 3.1 Publication Channels (MQ1)

Table 3 lists the journals and conferences/symposiums in which the selected papers of this systematic mapping study were published. Conferences/symposiums are the main target venue of studies on MV techniques in SE datasets with 19 papers (54 %) whereas only 12 articles (34 %) were published in journals. A great diversity could be found when analyzing the publication channels of the 35 articles.

## 3.2 Publication Trend (MQ2)

Figure 2 presents the number of articles published per year from 2000 to 2013. The earliest papers in our pool were published in 2000 [6, 7]. According to [29, 33] carried out the first study on the effect of missing data techniques on software development effort prediction. When analyzing the articles' distribution over time

**Table 3** Datasets used for MV techniques evaluation

| Publication venue | Type | # studies | Percent (%) |
|---|---|---|---|
| Journal of Systems and Software (JSS) | Journal | 5 | 14.41 |
| International Symposium on Software Metrics | Symposium | 4 | 11.41 |
| Software Engineering, IEEE | Journal | 3 | 8.57 |
| Empirical Software Engineering | Journal | 2 | 5.70 |
| International Conference on Predictive Models in Software Engineering (PROMISE) | Conference | 2 | 5.70 |
| International Conference on Software Engineering (ICSE) | Conference | 2 | 5.70 |
| Other conferences/symposium | | 11 | 31.40 |
| Other journals | | 2 | 5.70 |
| Others | | 4 | 11.41 |
| Total | | 35 | 100 |

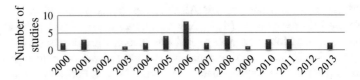

**Fig. 2** Publication per year

(Fig. 2), we noticed that the trends of MV publication in SE data sets are characterized by discontinuity. In fact, no papers were published in 2002 and 2012. Besides, from 2005 to 2008, 18 papers were published (around 51 % of selected papers) which means that this research topic has gained an increased attention during those four years but it decreased afterwards.

## 3.3 Research Types (MQ3) and Research Approaches (MQ4)

Regarding the research types, Fig. 3 shows that 94 % of the papers (33) presented evaluations of MV techniques and 40 % of the selected studies (14) presented solutions (new MV technique or a significant extension of an existing technique) while two papers were classified as 'other'. Figure 3 also shows that the most adopted research approach is history-based evaluation (94 % of the papers). Table 4 shows that the most used dataset is ISBSG database with 17 evaluations (33 %), followed by domain specific data with 11 evaluations (23 %). [7] And [6]

**Fig. 3** Research types and research approaches

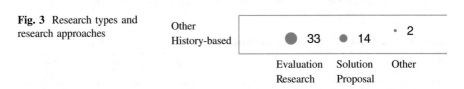

**Table 4** Publication venues and distribution of selected studies

| Datasets | # studies | # projects | Percent (%) | Source |
|---|---|---|---|---|
| ISBSG | 17 | >1000 | 33 | [12] |
| Domain specific data | 12 | | 23 | |
| Experience (Finnish) | 6 | 622 | 11 | [29] |
| CCCS | 6 | 282 | 11 | [8] |
| Desharnais | 4 | 81 | 8 | [33] |
| Kemerer | 3 | 15 | 6 | [33] |
| JM1-2445 (NASA) | 2 | 2445 | 4 | [8] |
| Bank data set | 2 | 18 | 4 | [33] |

were classified as 'other' since their primary purpose was not to evaluate or give a solution to the MV problem in SE context, but they used multiple imputations as a solution to the missing data they encountered.

## 3.4 MV Techniques (MQ5)

### 3.4.1 MV Techniques Distribution

Figure 4 shows the frequency of the MV techniques in selected papers and reveals that 97 % (34 papers) studied imputation techniques, and 29 % (10 papers) discussed deletion methods, while toleration methods only appear in 11 % (4 studies) of total selected papers. Toleration techniques were discussed in four papers, and were mainly used in comparison with imputation techniques [4, 13, 28, 39]. Even if deletion techniques are the most appealing to practitioners due to their simplicity, they have many disadvantages following the elimination of valuable data such as the loss of precision and bias of results [2, 31]. Imputation methods are especially useful in situations where a complete data set is required for the analysis or the mechanism causing the missing data is not random [29]. In fact, some data analysis methods can be only applied with complete data set. In this case, deletion or toleration techniques cannot be used. From the above, we note that deletion techniques make small software engineering data sets even smaller; while toleration techniques cannot provide complete data sets for analysis; hence, only imputation techniques can be used for the purpose of completing software data sets which may explain why deletion and toleration techniques are rarely used in the field of software engineering and why they have been often replaced with imputation methods.

The issue we encountered during data extraction at this level was that some studies used different terminologies for the same MV imputation technique. For example, some studies used instance based learning as the synonym of KNN. Similarly, some studies used regression trees, but other ones used classification and regression trees; both techniques belong to the category of decision trees. To avoid ambiguity, we adopted the terms KNNI and decision trees uniformly in data extraction and throughout this study. Note that when new methods are proposed by combining existing ones, each method is counted separately as in [22] which proposed a combination of SVDI and KNNI. Results show that 26 studies discussed non-ML imputation techniques while 24 studies discussed ML imputation techniques. Figure 5 shows that in the beginning of their exploration of MV techniques in SE datasets, researchers tended to investigate non-ML techniques. However, we

**Fig. 4** Distribution of MV methods

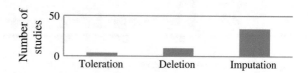

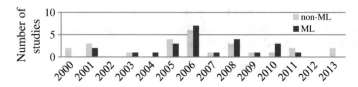

**Fig. 5** Distribution of ML and non-ML imputation techniques over years

notice that in 2004, 2006, 2008 and 2010, ML techniques were investigated more that non-ML techniques. But, non-ML techniques regained attention in the last three years. This can be explained by the difficulty of handling ML techniques and the required knowledge of other fields such as artificial intelligence and data mining instead of non-ML techniques which are appealing to researchers due to their simplicity to implement.

### 3.4.2 ML Imputation Techniques Distribution

Machine learning (ML) algorithms have proven to be of great practical value in a variety of application domains. Therefore, many researchers applied ML techniques to the field of MV imputation in SE datasets. Figure 6 shows distribution of ML techniques used to impute missing values in SE datasets. We concluded that KNNI was by far the most used: it was cited in 63 % (22 papers) of total selected papers, followed by DTI and FCI with equally 11 % (4 studies) of selected papers. As mentioned earlier, the most investigated ML technique was KNNI; its performance was analyzed in different contexts, and proven to be efficient and stable [4]. The most important remark about ML imputation techniques in SE datasets is their lack of variety and diversity. Except for KNNI, no ML imputation method was cited more than 4 times and it seems that researchers tended to lean toward KNNI in most of papers.

### 3.4.3 Non-ML Imputation Techniques Distribution

Figure 7 shows the distribution of non-ML imputation techniques: MEI and MI are the most used methods. Indeed MEI was investigated in 19 articles (54 %) and MI in 15 articles (43 %). MEI has the advantage of being extremely simple [27]. On the

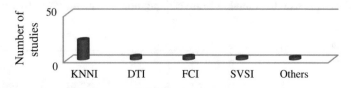

**Fig. 6** Distribution of ML imputation techniques

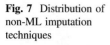

**Fig. 7** Distribution of non-ML imputation techniques

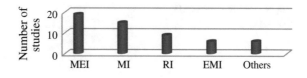

**Fig. 8** Distribution of data types

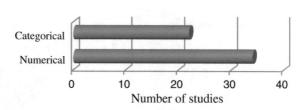

other hand, MI is an attractive choice as a solution to missing data problems as it represents a good balance between quality of results and ease of use [2]. In contrast to ML techniques, the percentages of studies investigating different non-ML techniques are close: there is not a big difference between them which reflects the diversity in using these imputation techniques.

## 3.5  Types of Data (MQ6)

Figure 8 shows that 97 % of selected papers (34) discussed numerical data, while only 63 % (22 papers) were dedicated to treating missing categorical data. Results show that even though the majority of software projects data sets contain mostly categorical attributes with many missing values, only few researchers have investigated categorical data imputation techniques [24]. Categorical data are more challenging because they are difficult to handle and the major issue is the lack of suitable imputation techniques for them. In fact, most of the techniques applied to this type of data are simple reuse techniques originally designed for numerical data.

## 3.6  Types of MV (MQ7)

Figure 9 shows that all selected studies treated MCAR mechanism, 46 % (16 studies) analyzed MAR mechanism, while MNAR mechanism appears in 37 % of the studies (13 studies). Moreover, results indicate that 54 % (19 studies) of selected papers treated only MCAR which means they assumed that data were missing randomly without any specific reason. The remaining papers compared the performances of MV techniques according to at least two of the missingness mechanisms (MCAR, MAR, and MNAR) of Sect. 2.5.

**Fig. 9** Distribution of MV
data types

**Fig. 10** Distribution of SE
subfields

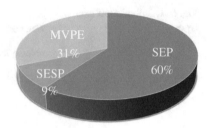

## 3.7 Objectives of Investigating MV Techniques in SE Datasets (MQ8)

Figure 10 shows that 60 % of the selected studies investigated missing data techniques to serve the software effort prediction models (SEP), 31 % of studies used MV techniques for performance comparison without predicting or evaluating any software project attribute (MVPE) and few selected studies (9 %) have used MV techniques in the context of predicting or evaluating a software project attribute other than effort (NESP) [6, 7, 9].

## 4 Implication for Research and Practice

The findings of this systematic mapping study have implications for researchers and practitioners working in the empirical software engineering field. It will allow them to find out the existing techniques and approaches to deal with MV in SE datasets. This study concluded that the trends of MV publications are characterized by discontinuity; therefore, researchers are encouraged to conduct more research on the use of MV methods in SE data sets. Most of the selected studies focused on the evaluation of existing MV techniques in SE datasets using historical databases and maybe researchers must turn their attention to developing new MV techniques with case study based evaluations. This study concluded that ML imputation techniques have not been sufficiently investigated. Therefore, researchers are encouraged to explore new techniques, ones that haven't been applied yet to SE context or haven't got much attention such as fuzzy logic, neural network or association rules. Regarding data types, researchers are encouraged to investigate categorical data, in order to implement imputation methods that are designed especially for this type of data since most of SE datasets contain a high number of categorical attributes. This study showed that most researchers do not take into consideration the missing mechanism when applying MV techniques in SE datasets. More research is needed to determine which MV techniques can be used for a specific missingness mechanism to obtain a higher performance.

# 5 Threats to Validity

In order to ensure that most relevant primary studies were being included keywords related to missing values as well as terms related to empirical SE were used in the search string to discover a wide range of papers covering MV techniques in SE datasets. However, many terms may have been missed in the search string which could have affected the final results of this paper. This issue would only have a minor influence since we used different libraries and scanned references of relevant papers in order to minimize the risk of missing any relevant paper. The QA process may be considered as subjective which could have affected the QA results. The final decision to select a study depended on the two authors who conducted the search process. If a disagreement arose between them, then a discussion took place until an agreement was reached. Two authors carried out the data extraction and classification of the primary studies, and reviewed the final results. The decision on the data to collect and how to classify the papers therefore depended on the judgment of the two first authors conducting this systematic mapping study. In order to present relevant results that can be exploited by other researchers, the search string, the databases and the inclusion/exclusion criteria and every step performed were presented in detail. In our opinion, slight differences based on publication selection bias and misclassification would not alter the main conclusions drawn from the 35 papers identified in this mapping study.

# 6 Conclusion and Future Work

The aim of this systematic mapping study was to examine the current research on the use of MV techniques in SE datasets by selecting 35 papers from a total of 288 candidate articles. These 35 selected studies were classified according to six classification criteria: research types, research approaches, MV techniques, data type, MV types, and objectives behind using MV techniques. In addition publication channels and trends were identified. The principal findings of our study were the following. (MQ1): The topic of MV techniques in SE data sets is taken seriously by researchers, as observed by the amount of publications in conferences/symposiums and journals. This study reveals that conferences/symposiums are the main target of publications. (MQ2): The timescale for included articles extends from 2000 to 2013 and the trends of MV publication in SE are characterized by discontinuity since no paper was published in 2002 and 2012. (MQ3): Most of the papers presented evaluations of MV techniques, solution proposals were presented as well while only two papers gave opinions about MVs techniques. (MQ4): Most of the selected studies belong to the history-based evaluation research approach and the most used dataset is the International Software Benchmarking Standard Group (ISBSG) database. (MQ5): The results reveal that imputation techniques were the most investigated, followed by deletion methods and then toleration techniques. (MQ6): This study showed a lack in exploration of missing categorical data treatment despite the huge amount of categorical data in software projects databases. (MQ7): The results indicate that 54 % of selected papers treated only MCAR, which means they assumed that data were missing randomly. However, most papers insist on taking the missing mechanisms into account while investigating MV techniques and show how these mechanisms can significantly affect their performance. (MQ8): The main motivation behind investigating MV techniques in SE datasets was to predict software development effort. This study could help practitioners to identify techniques with which to enhance the MV treatment in their projects, and it may also help researchers to identify both the datasets to be used in the evaluation of their studies and channels in which to publish their research results.

# References

1. Abran, A., Moore, J.W.: Guide to the software engineering body of knowledge (SWEBOK). IEEE Computer Society (2004)
2. Bala, A.: Impact analysis of a multiple imputation technique for handling missing value in the ISBSG repository of software projects, Ph.D thesis, supervised by Alain Abran, Ecole de technologie superieure—ETS, University of Quebec, Montreal Canada (2013)
3. Calikli, G., Bener, A.: An algorithmic approach to missing data problem in modeling human aspects in software development. In: PROMISE'13 Proceedings of 9th International Conference on Predictive Models in Software Engineering, Article No. 10 (2013)
4. Cartwright, M., Shepperd, M.J., Song, Q.: Dealing with missing software project data. In: Proceedings 9th IEEE International Software Metrics Symposium (Metrics'03), Sydney, Australia, pp. 154–165 (2003)
5. Chan, V.K.Y., Wong, W.E., Xie, T.F.: Application of a statistical methodology to simplify software quality metric models constructed using incomplete data samples. In: Sixth International Conference on Quality Software (2006)
6. El-Emam, K., Birk, A.: Validating the ISO/IEC 15504 measures of software development process capability. J. Syst. Softw. **51**(2), 119–149 (2000)
7. El-Emam, K., Birk, A.: Validating the ISO/IEC 15504 measure of software requirements analysis process capability. IEEE Trans. Softw. Eng. **26**(6), 541–566 (2000)
8. Hulse, J.V., Khoshgoftaar, T.M.: Incomplete-case nearest neighbor imputation in software measurement data. Inf. Sci. **259**, 630–637 (2001)
9. Hulse, J.V., Khoshgoftaar, T.M.: A comprehensive empirical evaluation of missing value imputation in noisy software measurement data. J. Syst. Softw. **81**(5), 691–708 (2008)
10. Hulse, J.V., Khoshgoftaar, T.M, Seiffert, C.: A comparison of software fault imputation. In: Procedures 5th International Conference on Machine Learning and Applications, ICMLA'06, 135–142 (2006)
11. Idri, A., Amazal, F., Abran, A.: Analogy-based software development effort estimation: a systematic mapping and review. Inf. Softw. Technol. **58**, 206–230 (2015)

12. ISBSG, Data R8. International Software Benchmarking Standards Group, www.isbsg.org, Oct 18 (2005)
13. Jingzhou, L., Al-Emran, A., Ruhe, G.: Impact analysis of missing values on the prediction accuracy of analogy-based software effort estimation method AQUA. In: First International Symposium on ESEM, pp. 126–135 (2007)
14. Jonsson, P., Wohlin, C.: An evaluation of k-nearest neighbour imputation using likert data. In: Proceedings 10th IEEE International Software Metrics Symposium (Metrics'04), Chicago, Illinois, pp. 108–118 (2004)
15. Jönsson, P., Wohlin, C.: Benchmarking k-nearest neighbour imputation with homogeneous Likert data. Empirical Software Engineering 11(3), 463–489 (2006)
16. Khoshgoftaar, T.M., Hulse, J.V.: Imputation techniques for multivariate missingness in software measurement data. Softw. Qual. J. 16(4), 563–600 (2008)
17. Khoshgoftaar, T.M., Folleco, A., Hulse, J.V, Bullard, L.: Software quality imputation in the presence of noisy data. In: IEEE International Conference on Information Reuse and Integration, pp. 484–489 (2006)
18. Liu, Q., Qian, W., Atanas, A.: Application of missing data approaches in software testing research. In: International Conference on Electronics, Communications and Control, pp. 4187–4191 (2011)
19. Mockus, A.: Missing Data in Software Engineering. Guide to Advanced Empirical Software Engineering, pp. 185–200 (2008)
20. Myrtveit, I., Stensrud, E., Olsson, U.: Assessing the benefits of imputing ERP projects with missing data. In: Proceedings Seventh International Software Metrics Symposium (2001)
21. Myrtveit, I., Stensrud, E., Olsson, U.H.: Analyzing data sets with missing data: an empirical evaluation of imputation methods and likelihood-based methods. IEEE Trans. Softw. Eng. 27 (11), 999–1013 (2001)
22. Olanrewaju, R.F., Ito, W.: Development of an imputation technique—INI for software metric database with incomplete data. In: 4th Student Conference on Research and Development, pp. 76–80 (2006)
23. Ouhbi, S., Idri, A., Fernández-Alemán, J. L., Toval, A.: Requirements engineering education: a systematic mapping study. J. Requirements Eng. 1–20 (2013)
24. Panagiotis, S., Lefteris, A.: Categorical missing data imputation for software cost estimation by multinomial logistic regression. J. Syst. Softw. 79, 404–4014. http://promise.site.uottawa.ca/SERepository (2006)
25. Seo, Y.S., Yoon, K.A., Bae D.H.: An empirical analysis of software effort estimation with outlier elimination. In: Proceedings 4th International workshop on Predictor models in software engineering.(PROMISE'08), pp. 25–32 (2008)
26. Song, Q., Shepperd, M.: A new imputation method for small software project data sets. J. Syst. Softw. 80(1), 51–62 (2007)
27. Song, Q., Shepperd, M., Cartwright, M.: A short note on safest default missingness mechanism assumptions. Empirical Softw. Eng. 10(2), 235–243 (2005)
28. Song, Q., Shepperd, M., Chen, X., Liu, J.: Can k-NN imputation improve the performance of C4.5 with small software project data sets? A comparative evaluation. J. Syst. Softw. 81(12), 2361–2370 (2008)
29. Strike, K., El Emam, K., Madhavji, N.: Software cost estimation with incomplete data. IEEE Trans. Softw. Eng. 27(10), 890–908 (2001)
30. Sumanth, Y.: An Empirical study of imputation techniques for software data sets. Thesis (PhD) Louisiana State University and The Department of Computer Science Agricultural and Mechanical College (2005)
31. Tamura, K., Kakimoto, T., Toda, K., Tsunoda, M., Monden, A., Matsumoto, K.: Empirical Evaluation of Missing Data Techniques for Effort Estimation. http://citeseerx.ist.psu.edu/viewdoc/summary?. doi:10.1.1.145.780 (2009)
32. Twala, B., Cartwright, M: Ensemble imputation methods for missing software engineering data. In: 11th IEEE International Symposium on Software Metrics, pp. 10–30 (2005)

33. Twala, B., Cartwright, M.: Ensemble missing data techniques for software effort prediction. Intell. Data Anal. 299–331 (2010)
34. Twala, B., Cartwright, M., Shepperd, M.: Comparison of various methods for handling incomplete data in software engineering databases. In: Proceedings of the 4th International Symposium on Empirical Software Engineering, Noosa Heads, Australia (2005)
35. Twala, B., Cartwright, M., Shepperd, M.: Ensemble of missing data techniques to improve software prediction accuracy. In: ICSE'06 Proc. 28th International Conference on Software Engineering, pp. 909–912 (2006)
36. Wong, W.E., Zhao, J., Chan, V.K.Y.: Applying statistical methodology to optimize and simplify software metric models with missing data. In: SAC'06 Proceedings 2006 ACM symposium on Applied computing, pp. 1728–1733 (2006)
37. Xie, T., Wong, W.E.: An improved method to simplify software metric models constructed with incomplete data samples. In: 7th International Conference on Fuzzy Systems and Knowledge Discovery, pp. 1682–1688 (2010)
38. Xie, T., Wong, W.E., Ding, W.: Simplifying software metric models via hierarchical LASSO with incomplete data samples. In: Second World Congress on Software Engineering (WCSE), vol. 2, pp. 161–164 (2010)
39. Zhang, W., Yang, Y., Wang, Q.: Handling missing data in software effort prediction with naive Bayes and EM algorithm. In: Promise'11 Proceedings 7th International Conference on Predictive Models in Software Engineering. doi:10.1145/2020390.202 (2011)

# Testing and Code Review Based Effort-Aware Bug Prediction Model

K. Muthukumaran, N.L. Bhanu Murthy, G. Karthik Reddy
and Prateek Talishetti

**Abstract** Incremental and iterative processes compel vendors to release numerous versions of the software and quality of each and every version is the top most priority for vendors. Bug prediction models attempt to rank or identify the bug prone files so that quality assurance team can optimally utilize their resources to ensure the quality delivery. Recently, Researchers proposed effort aware bug prediction models wherein files are ranked not only based on their bug proneness but also on the effort that is required to perform quality assurance activities like testing, code review etc. on bug prone files. And Cyclomatic Complexity is being considered as a metric for the effort required irrespective of the assurance activity that will be undertaken during post prediction phases. We propose effort aware bug prediction model that will consider the effort required to perform the specific quality assurance activity and empirically prove the supremacy of these models over the existing effort aware models.

**Keywords** Defect prediction · Evaluation · Effort estimation · Cost-benefits

K. Muthukumaran · N.L. Bhanu Murthy (✉) · P. Talishetti
BITS Pilani Hyderabad Campus, Hyderabad, India
e-mail: p2011415@hyderabad.bits-pilani.ac.in

K. Muthukumaran
e-mail: p2011415@hyderabad.bits-pilani.ac.in

P. Talishetti
e-mail: h2011013@hyderabad.bits-pilani.ac.in

G. Karthik Reddy
Stony Brook University, New York, USA
e-mail: kgreddy@cs.stonybrook.edu

© Springer International Publishing Switzerland 2016
R. Lee (ed.), *Software Engineering, Artificial Intelligence, Networking and Parallel/Distributed Computing*, Studies in Computational Intelligence 653, DOI 10.1007/978-3-319-33810-1_2

# 1   Introduction

Quality has always been a non-compromising priority task for all IT companies and post release bugs or bugs experienced by end customer hampers the quality heavily. Quality assurance activities, such as tests or code reviews, are an expensive, but vital part of the software development process. Any support that makes this phase more effective may thus improve software quality and reduce development costs. With time and manpower being limited or gets limited for quality assurance activities, it would be great if one can predict bug prone files in early stages of software development so that the project team can focus more on those bug prone files. Also, it helps the team to allocate or manage their critical resources well. It has been observed that the distribution of defects follows a Pareto-principle, that is, that most of the bugs are located in only few files [7]. Hence the research on bug prediction models have become significant and more than hundred research papers have been published.

The project team might perform quality assurance activities on bug prone files to catch bugs earlier to release of the software. The quality assurance activities could be peer code reviews, expert code reviews, testing etc. The team can choose one or more quality assurance activities that are apt for the project so that maximum number of defects will be uncovered. Also, project team has to allocate resources/revenues to complete these activities. Though it is ideal to get assurance activities done for all of the bug prone files, it might not be feasible due to cost implications. Hence project team would act on top $x\%$ of bug prone files based on the costs that can be borne by the team.

The top $x\%$ of files will be selected by sorting files in the descending order of number of predicted bugs and this approach has been adopted by researchers Ohlsson and Alberg [16], Khoshgoftaar and Allen [9]. We illustrate the limitations of this approach with the following example. Generally it is assumed that after performing the quality assurance activities like code review of files, all bugs in the file will be uncovered. Suppose project team would like to take up code review of top $x\%$ of files and there are two files File A, File B with the following characteristics.

| File | LOC | # of predicted bugs | Effort per bug |
|------|-----|---------------------|----------------|
| File A | 100 | 2 | 50 |
| File B | 1200 | 4 | 300 |

File B appears prior to File A as per the above mentioned sorted order and File A might have been ignored when project team selects top $x\%$ of files. The Effort per bug indicates the effort required to uncover one bug and is defined as the ratio of LOC and the number of bugs. The effort per bug for File A (50) is quite low as compared to File B (300) whereas the classical ordering ranks File B on top of File A. This issue has been addressed by Mende and Koschke [13].

Mende and Koschke [13] proposed a strategy to include the notion of effort into defect prediction models. They propose to rank files with respect to their effort per bug. They use McCabe's cyclomatic complexity as the surrogate measure for effort. They evaluate their new model against the naive model which just ranks files on the predicted number of bugs, by adopting to effort aware measures like CE [1]. They apply Demšar's non-parametric tests to conclude that their new model is significantly cost effective than the naive model.

**Contributions**: In this work we argue that the measure of effort should not be a generic measure such as cyclomatic complexity but instead it should be one that is specific to the kind of activity involved in the quality assurance process. We identify two most popular quality assurance activities namely code review and unit testing. We use lines of code to measure the effort involved in code reviews and the number of test cases to measure the effort in case of unit testing. We compare the cost effectiveness of our specific effort based models to generic effort based models.

## 2 Related Work

Many approaches in the literature use the source code metrics and change metrics as predictors of bugs. In this paper we have used the CK metrics suite [5], object oriented metrics, lines of code and some change metrics as features to build the bug prediction model. Basili et al. [3], Tang et al. [19], Cartwright and Shepperd [4], Subramanyam and Krishnan [18] explored the relationship between CK metrics and and defect proneness of files. Basili et al. [3] found that WMC, RFC, CBO, DIT and NOC are correlated with faults while LCOM is not associated with faults based on experiments on eight student projects. Tang et al. [19] validated CK metrics using three industrial real time systems and suggest that WMC and RFC can be good indicators of bugs. Cartwright and Shepperd [4] found DIT and NOC as fault influencers. Subramanyam and Krishnan [18] investigated the relationship between a subset of CK metrics and the quality of software measured in terms of defects. Though they proved the association, they conclude that this observation is not consistent with different OO languages like C++ and Java.

Nagappan et al. used a catalog of source code metrics to predict post release defects at the module level on five Microsoft systems [15]. Ostrand et al. [17] conclude that lines of code is a significant influence on the faults. Also their simple model based only on lines of code achieves roughly 73 % bugs in the top 20 % of files which is only 10 % less than their full model. Krishnan et al. [10] studied the performance of a large set of change metrics on seven versions of multiple Eclipse products between the years 2002–2010. They conclude that, the performance of change metrics improved for each product as it evolved across releases. For each product, they identified a small, stable set of change metrics that remained prominent defect predictors as the products evolved. For all the products, across all the releases, the change metrics "maximum changeset", "number of revisions" and "number of authors" were the good predictors.

Ohlsson and Alberg [16], Ostrand et al. [17] advocate models which rank files based on their fault proneness. Khoshgoftaar and Allen [9] coined the term Module-Order-Model (MOM). It enables to select a fixed percentage of modules for further treatment which is a more realistic scenario for projects with a fixed quality assurance budget. MOMs can be evaluated by assessing the percentage of defects detected at fixed percentages of modules. Ostrand et al. [17] propose a model which found up to 83 % of the defects in 20 % of the files. One way to graphically evaluate MOMs are lift charts, sometimes known as Alberg diagrams [16]. Ostrand et al. validates Pareto principle in bug prediction and their results show that 20 % of files have 84 % of defects. Ostrand et al. [17] conclude that a prediction using defect densities is able to find more defects in a fixed percentage of code, but argue that the testing costs are, at least for system tests, not related to the size of a file.

Mende and Koschke [12] compare the performance of various classifiers over lines of code based lift charts. They conclude that performance measures should always take into account the percentage of source code predicted as defective. In their subsequent work [13], they propose two strategies to incorporate the treatment effort into defect prediction models. But both their strategies include cyclomatic complexity as the effort for quality assurance. In this work we propose that specific effort measures perform better than generic effort based classifiers.

## 3   Metrics and Datasets

We have considered a hybrid of source code metrics which include the Chidamber and Kemerer metric suite [5], few object oriented metrics and change metrics. Table 1 describes details about metrics considered. We need to have the number of test cases for each source file to achieve the objectives of this work. The benchmarked open source datasets, that are used for bug prediction research, do not have the number of test cases against each file. Hence we, acquired data from a private software company for three projects in which a record of all testing activities for every file is recorded. For the sake of convenience and anonymity, let us consider them as datasetA, datasetB and datasetC. The details about the number of files and number of defects in each dataset are shown in Table 2.

## 4   Experiment and Results

We build defect prediction models using the metrics mentioned in Table 1 for each of the three datasets in Table 2. We use stepwise linear regression algorithm to build our model and predict the number of faults for a file. Regression algorithm takes the number of faults into consideration while building the model whereas a classification algorithm such as Naive Bayes classifier only considers the label of the file. Since we intend to predict effort per bug of a file we use stepwise linear

**Table 1** Metrics

| Source code metrics | |
|---|---|
| WMC | Weighted method count |
| DIT | Depth of inheritance |
| NOC | Number of children |
| CBO | Coupling between object classes |
| RFC | Response for a class |
| LCOM | Lack of cohesion in methods |
| CE | Efferent coupling |
| NPM | Number of public methods |
| CC | Cyclomatic complexity is a popular procedural software metric equal to the number of decisions that can be taken in a procedure |
| NOF | Number of fields in the class |
| NOI | Number of interfaces implemented by the class |
| LOC | Number of lines of code in the file |
| NOM | Number of methods in the class |
| Fan-in | Number of other types this class is using |
| Fan-out | Number of other types using this class |
| PC | Percentage of the file commented |
| *Change code metrics* | |
| noOfBugs | Number of bugs found and fixed during development |
| noOfStories | Number of stories this file is part of |
| noOfSprints | Number of sprints this file is part of |
| revisionCount | Number of commits this file is involved in |
| noOfAuthors | Number of authors who worked on this file during the release |
| LocA | Number of lines of code added in total to this file during development |
| MaxLocA | Max number of lines of code added among all commits to this file during development |
| AvgLocA | Average number of lines of code added among all commits to this file during development |
| LocD | Total number of lines of code deleted among all commits to this file during development |
| MaxLocD | Max number of lines of code deleted among all commits to this file during development |
| AvgLocD | Average number of lines of code deleted among all commits to this file during development |
| LocAD | Total number of lines of code added-deleted among all commits to this file during development |
| MaxLocAD | Max number of lines of code added-deleted among all commits to this file during development |
| AvgLocAD | Average number of lines of code added-deleted among all commits to this file during development |

**Table 2** Datasets

| Projects | Number of files | Post release defects |
|----------|-----------------|----------------------|
| DatasetA | 2257 | 252 |
| DatasetB | 2557 | 652 |
| DatasetC | 2151 | 349 |

regression algorithm. Nagappan and Ball [14] also used stepwise linear regression to predict file defect density. We adopt the method of building bug prediction models using 2/3rd data and testing the models with remaining 1/3rd data. The division of training and testing data is being done randomly. The data is split randomly fifty times and the average of values were recorded as final results in experiments conducted for this work.

Khoshgoftaar and Allen [9] proposed software quality model, MOM (Module Order Model) that is generally used to predict the rank-order of modules or files according to a quality factor [9]. We take this quality factor to be number of defects in our study. We apply stepwise linear regression algorithm and order files as per their predicted value of bugs. MOMs can be evaluated by assessing the percentage of defects detected against the fixed percentages of files. One way to graphically evaluate MOMs are lift charts, sometimes known as Alberg diagrams [16]. They are created by ordering files according to the score assigned by a prediction model and denoting for each fixed percentage of files on the x-axis, the cumulative percentage of defects identified, on the y-axis. Thus for any selected percentage of files, one can easily identify the percentage of predicted defects. We have drawn Alberg Diagrams [16] for two module-order models namely actual and simple prediction for each of the project and the plots are depicted in Fig. 1a–c.

The optimal or ideal or actual model (represented as red curve in the above plot) is drawn by arranging the modules in the decreasing order of actual defects. The simple prediction model(represented as blue curve in the plots) is drawn by arranging the files in the decreasing order of predicted number of faults. The predicted number of faults is value outputted by the stepwise linear regression algorithm for each file. The optimal or ideal model represents the best possible model that any defect prediction model can aspire for. The goodness of any prediction model depends on its closeness to the optimal curve, closer to the optimal curve better the prediction model is. And the major advantage of this method is that the model is flexible enough to predict bugs depending on the costs that can be borne by project team towards quality assurance activities.

The Pareto principle for this problem is '20 % of files should have 80 % of bugs' and we have evaluated the number of bugs that can be found in top 20 % of files for the prediction models across all the datasets and these values are tabulated in Table 3. The percentage of bugs found using these models across projects is found to be varying between 39.56 and 80 %. These results are depicted visually in Fig. 1a–c. Now, we shall explain the results of datasetB. The results, suggest that if we do quality assurance activities on the 20 % of files we can catch 80 % of the bugs. Although this is an excellent result, we shall now explore the amount of effort

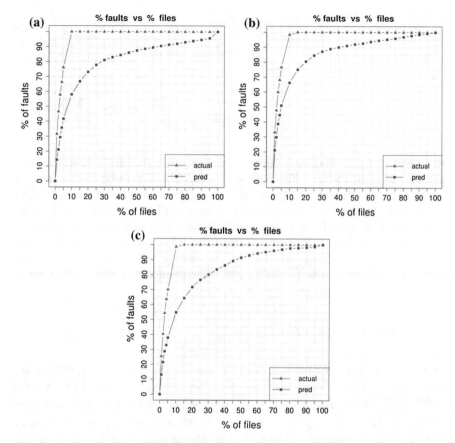

**Fig. 1** Actual and simple prediction. **a** DatasetA. **b** DatasetB. **c** DatasetC

**Table 3** Percentage of bugs in top 20 % of files

| Projects | Actual | Predicted |
|----------|--------|-----------|
| DatasetA | 100 | 73.18 |
| DatasetB | 100 | 80.48 |
| DatasetC | 100 | 71.72 |

it takes to review/test these files. In practice there are couple of quality assurance activities that can be performed soon after the bugs are predicted through the learning model. We consider two such popular quality assurance activities namely code review and unit testing.

We assume that effort to do code review is directly proportional to the lines of code one has to review and for unit testing, effort is proportional to the number of test cases one has to perform manually. The effort required to conduct the two quality assurance activities for the top 20 % files mentioned in Table 3 are tabulated in Table 4. From these results, we can infer the following: for datasetB, by

Table 4 Percentage of effort required for the top 20 % of files

| Projects | Model | % Lines of code | % Testcases |
|----------|-----------|-----------------|-------------|
| DatasetA | Actual | 39.75 | 36.67 |
|          | Predicted | 66.15 | 78.06 |
| DatasetB | Actual | 47.76 | 56.97 |
|          | Predicted | 64.72 | 76.91 |
| DatasetC | Actual | 41.66 | 40.05 |
|          | Predicted | 63.81 | 66.4 |

reviewing the top 20 % of the predicted files, one can uncover 80 % of the bugs and the effort required for these files is reviewing 64.72 % of the total lines of code or performing 76.91 % of the testcases. This shows that, although the number of files predicted as bug prone is only 20 %, the amount of effort required to uncover the bugs from these files is very large which is only slightly better than a random inspection model which gives n% of bugs for n% of effort. This result is also observed in the remaining two datasets. In view of effort, the standard Pareto principle should be modified as follows: '20 % of effort should give 80 % of bugs'.

## 4.1 Effort Based Evaluation

In-order to minimize the effort involved to uncover the bugs in files, Mende and Koschke [13] propose a new ranking order. They rank the files based on the ratio: $\frac{predicted\ number\ of\ bugs}{effort}$ instead of predicted number of bugs alone. They use cyclomatic complexity as the surrogate measure for effort with the assumption that the lines of code as well as testcases correlate highly with cyclomatic complexity. In our datasets we have found the previous statement to be not true. Although the cyclomatic complexity is correlating highly with the lines of code, but it is not so with testcases. The Spearman correlations between cyclomatic complexity and lines of code, cyclomatic complexity and test cases is tabulated in Table 5.

Thus, instead of using a generic measure of effort such as cyclomatic complexity, we use lines of code and the number of testcases as our measures of effort. The predicted number of bugs term in the ratio is the value outputted by any learning algorithm which is in this case stepwise linear regression algorithm. The effort term in the ratio can be substituted with the lines of code, if quality assurance activity is code review or it can be number of testcases, if quality assurance activity is unit

Table 5 Spearman correlation

| Projects | Cyclomatic complexity and lines of code | Cyclomatic complexity and number of testcases |
|----------|------------------------------------------|-----------------------------------------------|
| DatasetA | 0.9387 | 0.2703 |
| DatasetB | 0.9379 | 0.5487 |
| DatasetC | 0.9358 | 0.3687 |

**Table 6** Percentage of defects caught by 20 % of effort (LOC)

| Projects | Actual | Simple prediction | $\frac{predicted\ \#bugs}{LOC}$ | $\frac{predicted\ \#bugs}{CC}$ |
|----------|--------|-------------------|------------------|-----------------|
| DatasetA | 70 | 22 | 33 | 30 |
| DatasetB | 65 | 35 | 42 | 40 |
| DatasetC | 60 | 25 | 40 | 38 |

**Table 7** Percentage of defects caught by 20 % of effort (testcases)

| Projects | Actual | Simple prediction | $\frac{predicted\ \#bugs}{TC}$ | $\frac{predicted\ \#bugs}{CC}$ |
|----------|--------|-------------------|------------------|-----------------|
| DatasetA | 59 | 10 | 66 | 6 |
| DatasetB | 50 | 24 | 33 | 14 |
| DatasetC | 64 | 16 | 55 | 6 |

testing. This ratio gives a higher rank to files which have less effort per bug and a lower rank to files which have more effort per bug. Using our proposed ranking orders, the percentage of defects captured by 20 % of effort are recorded in Tables 6 and 7. The results are depicted visually by a plot of % of effort versus % defects in Figs. 2a, b, 3a, b and 4a, b. Let us consider the Fig. 3a, b which correspond to datasetB. In both figures, the plots of $\frac{predicted\ number\ of\ bugs}{lines\ of\ code}$ and $\frac{predicted\ number\ of\ bugs}{testcases}$ is always higher than the plot of simple prediction. This indicates that, to uncover a certain amount of bugs, our proposed ranking orders require less effort than the traditional simple prediction rank order. Although the simple prediction model is better than our proposed model when considering only the % of bugs caught versus the % of files (Fig. 1a), it fails to perform better when considering the effort required to uncover the bugs (Fig. 3a, b). This result is observed in all the three datasets. Also, common to all the three datasets is the observation that the specific effort based

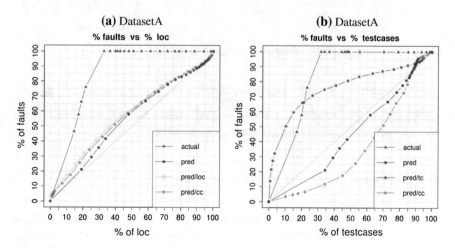

**Fig. 2** Effort-aware prediction: DatasetA. **a** DatasetA. **b** DatasetA

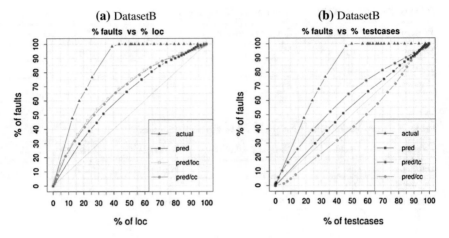

**Fig. 3** Effort-aware Prediction: DatasetB. **a** DatasetB. **b** DatasetB

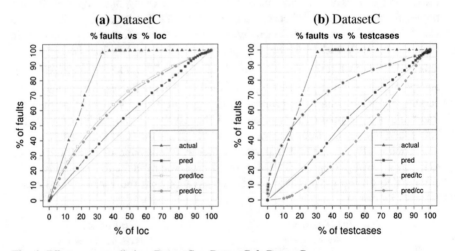

**Fig. 4** Effort-aware prediction: DatasetC. **a** DatasetC. **b** DatasetC

models i.e. $\frac{predicted\,number\,of\,bugs}{lines\,of\,code}$ and $\frac{predicted\,number\,of\,bugs}{testcases}$ are performing better than the $\frac{predicted\,number\,of\,bugs}{cyclomatic\,complexity}$. In order to quantify the performance of each classifier we consider an evaluation scheme described in the next section.

## 4.2 Percentage of Cost Effectiveness

Arisholm et al. [2] proposed an effort evaluation measure CE which stands for Cost Effectiveness. They calculate a cost effectiveness estimation based on the assumption that a random selection of n% of source lines contains n% of the defects.

A defect prediction model is cost effective only when the files predicted as defective contain a larger percentage of defects than their percentage of lines of code. Their performance measure CE can be obtained by calculating the area under the prediction model's curve which lies above a line of slope one. Mende and Koschke [12] defined the measure $p_{opt}$ which measures how close a model is to the ideal curve. Closer the curve to the ideal curve, the higher the value of $p_{opt}$. This measure takes into consideration the effort as well as the actual distribution of faults by benchmarking against a theoretically optimal model.

We consider the evaluation measure POA, Percentage of Area. This measure attempts to combine the aspects of both CE and $p_{opt}$, which has information about the lower bound of cost effectiveness, the random model and the upper bound, the theoretically optimal value. POA is the ratio of the CE of a model to the CE of the theoretically optimal model. This value gives the fraction of cost effectiveness a model achieves out of the theoretically maximum possible cost effectiveness.

$$POA = \frac{CE \ of \ model}{CE \ of \ the \ ideal \ model}$$

We present the POA values for the models when effort is lines of code in Table 8 and when effort is testcase in Table 9. From the results it is clear that specific effort based models outperform generic effort based measure cyclomatic complexity.

## 4.3 Statistical Significance

We now test whether the differences between the three models are statistically significant using the non parametric tests mentioned by Demšar [6]. Demšar uses

**Table 8** POA—lines of code

| Model | DatasetA | DatasetB | DatasetC |
|---|---|---|---|
| Ideal | 100 | 100 | 100 |
| Simple prediction | 14.23 | 35.59 | 16.03 |
| Predicted Number of Bugs / Lines of Code | 22.61 | 52.2 | 46.55 |
| Predicted Number of Bugs / Cyclomatic Complexity | 19.56 | 49.84 | 41.73 |
| Random model | 0 | 0 | 0 |

**Table 9** POA—testcases

| Model | DatasetA | DatasetB | DatasetC |
|---|---|---|---|
| Ideal | 100 | 100 | 100 |
| Simple prediction | 0.75 | 13.18 | 7.06 |
| Predicted Number of Bugs / Testcases | 75.13 | 38.65 | 67.22 |
| Predicted Number of Bugs / Cyclomatic Complexity | 0.73 | 0.3 | 0.19 |
| Random model | 0 | 0 | 0 |

the Friedman test [8] to check whether the null hypothesis, i.e. that all models perform equal on the datasets, can be rejected. It is calculated as follows:

$$\chi_F^2 = \frac{12N}{k(k+1)} \left( \sum_j R_j^2 - \frac{k(k+1)^2}{4} \right)$$

$$F_F = \frac{(N-1)\chi_F^2}{N(k-1) - \chi_F^2}$$

where $k$ denotes the number of models, $N$ the number of datasets, and $R_j$ the average rank of model $j$ on all data sets. $F_F$ is distributed according to the F-Distribution with $k - 1$ and $(k - 1)(N - 1)$ degrees of freedom. Once computed, we can check $F_F$ against critical values for the F-Distribution and then accept or reject the null hypothesis. When the Friedman test rejects the null hypothesis, we can use the Nemenyi post hoc test to check whether the difference of performance between two models is statistically significant. The test uses the average ranks of each model and checks for each pair of models whether the difference in their average ranks is greater than the critical difference CD.

$$CD = q_\alpha \sqrt{\frac{k(k+1)}{6N}}$$

where k and N are number of models and datasets respectively. $q_\alpha$ is a critical value which are based on the Studentized range statistic divided by $\sqrt{2}$. The Studentized range statistic depends on the alpha value which in this cases we take as 0.05 and the number of models $k$. For our setup, we used $k = 3$, $\alpha = 0.05$ and $q_\alpha = 2.85$.

We use Lessmann et al. [11] modified version of Demšar's significance diagrams to depict the results of Nemenyi's post hoc test: For each classifier on the y-axis, the average rank across the datasets is plotted on the x-axis, along with a line segment whose length encodes CD. All classifiers that do overlap in this plot do not perform significantly different and those that do not overlap, perform significantly different. The Nemenyi's post hoc significance plots are presented in Fig. 5a, b.

Consider Fig. 5a: $\frac{predicted\,no.\,of\,bugs}{lines\,of\,code}$ ranks the best, followed by $\frac{predicted\,no.\,of\,bugs}{cyclomatic\,complexity}$ and predicted no. of bugs. Although, the $\frac{predicted\,no.\,of\,bugs}{lines\,of\,code}$ performs better, the difference between it and $\frac{predicted\,no.\,of\,bugs}{cyclomatic\,complexity}$ is not significant.

In Fig. 5b: $\frac{predicted\,no.\,of\,bugs}{test\,cases}$ ranks the best, followed by predicted no. of bugs and $\frac{predicted\,no.\,of\,bugs}{cyclomatic\,complexity}$, $\frac{predicted\,no.\,of\,bugs}{lines\,of\,code}$ model performs significantly better than $\frac{predicted\,no.\,of\,bugs}{cyclomatic\,complexity}$.

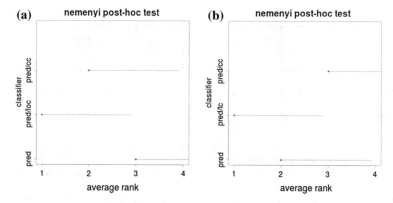

**Fig. 5** Nemenyi diagram for LOC and testcases. **a** Nemenyi Plot - Lines of Code. **b** Nemenyi Plot - Testcases

## 4.4 Threats to Validity

Since open source softwares are not developed in controlled environment and unavailability of proper testcases and considering the difficulties of linking them with appropriate files, We decided to do the experiments in proprietary software datasets. Similar experiments in different datasets may not yield similar results.

## 5 Conclusion

The consistent and repetitive results of three projects and generic Demšar test indicate us that the testing based effort aware models perform significantly better than generic effort aware models. The code review based effort aware model performs slightly better than generic model individually for each project though Demšar test show that there is no significant improvement. Hence, We conclude that effort aware models are sensitive to the type of quality assurance activity one undertakes and the effort required for these activities should be considered while building prediction models.

## References

1. Arisholm, E., Briand, L.C.: Predicting fault-prone components in a java legacy system. In: Proceedings of the 2006 ACM/IEEE International Symposium on Empirical Software Engineering, pp. 8–17. ACM (2006)
2. Arisholm, E., Briand, L.C., Johannessen, E.B.: A systematic and comprehensive investigation of methods to build and evaluate fault prediction models. J. Syst. Softw. **83**(1), 2–17 (2010)

3. Basili, V.R., Briand, L.C., Melo, W.L.: A validation of object-oriented design metrics as quality indicators. IEEE Trans. Softw. Eng. **22**(10), 751–761 (1996)
4. Cartwright, M., Shepperd, M.: An empirical investigation of an object-oriented software system. IEEE Trans. Softw. Eng. **26**(8), 786–796 (2000)
5. Chidamber, S.R., Kemerer, C.F.: A metrics suite for object oriented design. IEEE Trans. Softw. Eng. **20**(6), 476–493 (1994)
6. Demšar, J.: Statistical comparisons of classifiers over multiple data sets. J. Mach. Learn. Res. **7**, 1–30 (2006)
7. Fenton, N.E., Ohlsson, N.: Quantitative analysis of faults and failures in a complex software system. IEEE Trans. Softw. Eng. **26**(8), 797–814 (2000)
8. Friedman, M.: The use of ranks to avoid the assumption of normality implicit in the analysis of variance. J. Am. Stat. Assoc. **32**(200), 675–701 (1937)
9. Khoshgoftaar, T.M., Allen, E.B.: Ordering fault-prone software modules. Softw. Qual. J. **11** (1), 19–37 (2003)
10. Krishnan, S., Strasburg, C., Lutz, R.R., Goševa-Popstojanova, K.: Are change metrics good predictors for an evolving software product line? In: Proceedings of the 7th International Conference on Predictive Models in Software Engineering, ACM, New York, NY, USA, Promise '11, pp. 7:1–7:10. doi:10.1145/2020390.2020397. http://doi.acm.org/10.1145/2020390.2020397 (2011)
11. Lessmann, S., Baesens, B., Mues, C., Pietsch, S.: Benchmarking classification models for software defect prediction: a proposed framework and novel findings. IEEE Trans. Softw. Eng. **34**(4), 485–496 (2008)
12. Mende, T., Koschke, R.: Revisiting the evaluation of defect prediction models. In: Proceedings of the 5th International Conference on Predictor Models in Software Engineering, p. 7. ACM (2009)
13. Mende, T., Koschke, R.: Effort-aware defect prediction models. In: Proceedings of the Software Maintenance and Reengineering (CSMR), 2010 14th European Conference on, pp. 107–116. IEEE (2010)
14. Nagappan, N., Ball, T.: Use of relative code churn measures to predict system defect density. In: Proceedings of the 27th International Conference on Software Engineering, 2005. ICSE 2005, pp. 284–292. IEEE (2005)
15. Nagappan, N., Ball, T., Zeller, A.: Mining metrics to predict component failures. In: Proceedings of the 28th International Conference on Software Engineering, pp. 452–461. ACM (2006)
16. Ohlsson, N., Alberg, H.: Predicting fault-prone software modules in telephone switches. IEEE Trans. Softw. Eng. **22**(12), 886–894 (1996)
17. Ostrand, T.J., Weyuker, E.J., Bell, R.M.: Predicting the location and number of faults in large software systems. IEEE Trans. Softw. Eng. **31**(4), 340–355 (2005)
18. Subramanyam, R., Krishnan, M.S.: Empirical analysis of ck metrics for object-oriented design complexity: Implications for software defects. IEEE Trans. Softw. Eng. **29**(4), 297–310 (2003)
19. Tang, M.H., Kao, M.H., Chen, M.H.: An empirical study on object-oriented metrics. In: Proceedings of the Sixth International Software Metrics Symposium, 1999, pp. 242–249. IEEE (1999)

# Improving the Efficiency of Schedulability Tests for Fixed Priority Preemptive Systems

Guangliang Yu, Mengfei Yang, Hejie Sun and Hong Jiang

**Abstract** Efficient schedulability tests are required to check the schedulability of real-time systems, especially for online use. Since exact tests are very time consuming, great efforts have been paid to search for more efficient tests. Sufficient tests have been widely studied for they are usually performance efficiency. If a task set is proved schedulable by the sufficient test, then the exact response time does not need to be calculated. In this paper, we discuss the way of how to find better initial values for sufficient schedulability tests. Four parametric initial values are proposed and each can be tuned using a delta-parameter to reduce the number of iterations needed by response time calculation. Extensive evaluations show that the schedulability tests found by our method need less runtime than state of the art approaches.

## 1 Introduction

Fixed priority preemptive scheduling is widely used for real-time systems on a uniprocessor. Schedulability tests are usually required to guarantee the schedulability of these systems and check the satisfaction of timing constraints [1, 2]. A schedulability test is referred to as sufficient if it guarantees all the task sets that have passed the test can always meet their deadlines. Similarly, a schedulability test

G. Yu (✉) · H. Sun · H. Jiang
Beijing Institute of Control Engineering, Beijing, China
e-mail: ygl_222@126.com

H. Sun
e-mail: sunhejie@hotmail.com

H. Jiang
e-mail: smartpite@126.com

M. Yang
China Academy of Space Technology, Beijing, China
e-mail: yangmf@bice.org.cn

is referred to as necessary if it guarantees all the task sets that have failed the test will definitely result in a deadline miss at some point of time. A schedulability test is referred to as exact if it is both sufficient and necessary.

Exact schedulability tests for fixed priority scheduling on a uniprocessor are mainly based on processor's demands [3] or tasks' response times [4, 5]. For tasks having deadlines less than or equal to their periods (i.e., constrained-deadlines), no polynomial-time schedulability tests are known [3, 6]. Therefore an exact schedulability test may not be adequate for online use and a suitable sufficient schedulability test is needed to improve efficiency. If a task set is schedulable according to the sufficient test, then the exact response times do not need to be calculated, which saves amount of time significantly. In this paper, we are interested to offer a way of discovering better initial values for schedulability tests in order to decrease the time spent to calculate response time bounds.

Davis et al. [7] addressed performance issues with exact response time analysis for preemptive systems with fixed priority. They discussed different initial values for exact tests and compared their efficiencies. Bini et al. [8] introduced a continuous response time upper bound which can be a sufficient schedulability test and greatly improve the performance. Later they also introduced a quadratic-time response time upper bound which is much tighter [9]. Fisher et al. [10] proposed two FPTASs (fully-polynomial time approximation schemes) for checking feasibility of static-priority tasks subjected to release jitters executing upon a uniprocessor platform. These tests can be done in polynomial time with respect to the task set size and an "accuracy" parameter $1/e$. Their ideas are to perform the exact calculation in only $k$ first steps ($k$ is defined according to $e$) and to use a linear function to approximate the request bound subsequently. This bound was later improved by Nguyen et al. [11]. Based on [10, 11], Lu et al. [12] proposed a tunable response time upper bound for fixed priority real-time systems.

Based on the works of [8, 7], we extend their results via the following two steps. Firstly we assume the initial values parameterized with a series of delta parameters; secondly we use discrete simulation to search for more efficient schedulability tests for various task sets by selecting different delta parameters. Unlike [10], we first perform the sufficient test and if the sufficient condition fails, then the exact one is used instead. Since the proposed parametric initial values are tunable, one can find better performance schedulability tests through adjusting parameters. In a word, we present a way of finding more efficient schedulability tests for various task sets scheduled under a fixed priority preemptive scheduler.

In this paper, we first introduce preliminaries including the task model, exact schedulability tests and basic notions. Then we give further discussion of initial values, propose four parametric initial values and show how they work. Finally, we compare the performance of proposed approach with existing ones and draw a conclusion.

## 2 Background

### 2.1 System Model

In this paper, we consider a real-time system composed of a set of $n$ periodic real-time tasks $\tau = (\tau_1, \tau_2, \ldots, \tau_n)$ on a uniprocessor under fixed priority preemptive scheduling. Each task is assigned a unique priority $i$, from 1 to $n$ (where $n$ is the lowest priority). Each task $\tau_i$ is characterized by worst-case execution time $C_i$, period $T_i$, relative deadline $D_i$, and release jitter $J_i$ (defined as the maximum time between task arriving and being released). We assume that the period and execution time of task $\tau_i$ satisfy that $T_i > 0$ and $0 < C_i \leq D_i \leq T_i (i = 1, \ldots, n)$, i.e., constrained deadline. The maximum blocking time $B_i$ represents the longest time task $\tau_i$ can be blocked waiting for a lower priority task to complete its execution of corresponding critical sections. The utilization factor of the task $\tau_i$ is $U_i = C_i/T_i$. The utilization factor of a task set, denoted by $U$, is the sum of the utilizations of the tasks in the set, i.e., $U = \sum_{i=1}^{n} U_i$.

We use the notations $hp(i)$ and $lp(i)$ to mean the set of tasks with priorities higher than $i$, and the set of tasks with priorities lower than $i$ respectively. A task's worst-case response time (WCRT) $R_i$ is the longest time from the task being released to it completing execution. A task is referred to as schedulable if its worst-case response time is less than or equal to its deadline minus its release jitter, i.e., $R_i \leq D_i - J_i$. A task set is referred to as schedulable if all of its tasks are schedulable. In this paper, the cost of the context switches between tasks and the cache related preemption delay are not considered.

### 2.2 Exact Schedulability Tests

A critical instant [13] for task $\tau_i$ is the time when $\tau_i$ is released simultaneously with requests from all tasks with higher priorities. Then it is subject to the maximum possible delay, i.e., its worst-case response time.

A priority level-$i$ busy period [14] is a time interval $[a, b]$ within which jobs of priority $i$ or higher are processed throughout $[a, b]$ but none of jobs with priority $i$ or higher are processed in $(a - \varepsilon, a)$ or $(b, b + \varepsilon)$ for arbitrarily small $\varepsilon > 0$.

The response time analysis is based on the study of the busy period initiated by the critical instant (Here we assume the critical instant is "$t = 0$"). The critical instant is the worst-case scenario of the task under analysis, leading to the longest busy period, containing its WCRT. It is known that the WCRT of $\tau_i$ is equal to the smallest $t$ satisfying the following equation:

$$R_i = \min\{t > 0 | t = C_i + B_i + \sum_{j=1}^{i-1} \left\lceil \frac{t + J_j}{T_j} \right\rceil C_j\}, \tag{1}$$

where the summation term represents the total interference due to invocations of higher priority tasks released strictly before the end of the task $\tau_i$. Since $t$ appears on both sides of the equation, the recurrence relation can be solved by the fix-point iteration. The iteration is also proved to terminate [4] in a finite number of steps if $\sum_{k=1}^{i-1} U_k < 1$. Iteration starts with an initial value $t = R_i^0$ and ends when either $t = R_i^{n+1} = R_i^n$ in which case the worst-case response time is given by $t = R_i^n$ or when $t = R_i^n > D_i - J_i$ in which case the task is not schedulable. Note that the superscript denotes the number of iterations.

In fact, the right side of the inner equation of (1) gives the cumulative demands on the processor made by $\tau_1$ to $\tau_i$ over $t \in [0, t]$ shown in (2):

$$W_i^d(t) = C_i + B_i + \sum_{j=1}^{i-1} \left\lceil \frac{t + J_j}{T_j} \right\rceil C_j. \tag{2}$$

In a priority level-$i$ busy period initiated by the critical instant, the demand is not met until $W_i^d(t) = t = R_i$. So for $t \in (0, R_i)$, we have $W_i^d(t) > t$. It means that for any $t > 0$, if $W_i^d(t) \le t$, then there must be $t \ge R_i$, i.e., $t$ is an upper bound of $R_i$. Therefore a periodic task $\tau_i$ is schedulable if we find some $t \in [0, D_i - J_i]$ satisfying $W_i^d(t) \le t$.

So the implementation of the schedulability test can be done by two ways. Finding a solution $R_i^{UB} \in [0, D_i - J_i]$ of Eq. (1) or finding a point $t \in [0, D_i - J_i]$ subject to $W_i^d(t) \le t$.

It was proved that (Theorem 1 in [15]): if any $t$ in the fix-point iteration of (1) is less than or equal to the smallest solution to it, then the fix-point iteration is guaranteed to converge at the smallest solution (provided that at least one solution exists). According to this theorem, all the points in the time interval $(0, R_i]$ will lead to the recurrence of (1) to converge at $R_i$ and the more closer to $R_i$, the less iteration it takes. So finding initial values close to $R_i$ for (1) can significantly reduce the computation time.

On the other hand, checking every time point in $[0, D_i - J_i]$ to determine the schedulability is impossible. But it was pointed out in [3] that we only need to check the condition $W_i^d(t) \le t$ at time $t \in S_i$ where

$$S_i = \left\{ kT_j \mid j = 1, 2, \ldots, i; \; k = 1, 2, \ldots, \left\lfloor \frac{T_i}{T_j} \right\rfloor \right\}. \tag{3}$$

Accordingly reducing the number of points to be checked will save the time spent to perform the test.

# 3   Schedulability Tests

In this section, we first give detailed discussions about the influence of different initial values for schedulability tests on the number of iterations of Eq. (1) as well as the precision of calculated response time upper bounds. Then we propose four parametric initial values for schedulability analysis including a series of parametric test points to check the processor demand.

## 3.1   Discussions of Initial Values

Basically there are two conditions of a task: schedulable or not schedulable, which we can't know until the schedulability test is done. So firstly we will assume task $\tau_i$ is in fact schedulable, i.e., $0 < R_i \leq D_i - J_i$, though we haven't known it yet before the test. In this case if we randomly select some initial value $t = R_i^0$ from the interval $(0, +\infty)$, there can be three possible scenarios shown as follows:

- The selected initial value lies in the interval $(0, R_i]$. In this condition the iteration will converge at $R_i$ which is less than $D_i - J_i$, so we can conclude that task $\tau_i$ is schedulable. Also the bigger the initial value, the less the number of iterations.
- The selected initial value lies in the interval $(R_i, D_i - J_i]$. In this condition, the first iteration of (1) is actually to compute the cumulative demands $W_i^d(t = R_i^0)$. If $W_i^d(R_i^0) \leq R_i^0$, then we know that $W_i^d(R_i^0)$ must be one of the response time upper bounds of task $\tau_i$, so it can be concluded that $R_i \leq W_i^d(R_i^0) \leq R_i^0 \leq D_i - J_i$. Therefore we conclude that task $\tau_i$ is schedulable and the smaller the initial value, the smaller the upper bound. However, if $W_i^d(R_i^0) > R_i^0$, the iteration of Eq. (1) will finally converge at one of the response time upper bound $R_i^{UB}$. If $R_i^{UB} \leq D_i - J_i$, then we can still conclude that task $\tau_i$ is schedulable. In this case we also have the smaller the initial value, the smaller the upper bound. If $R_i^{UB} > D_i - J_i$, then nothing can be concluded. And also the number of iterations is undecided.
- The selected initial value lies in the interval $(D_i - J_i, +\infty)$. In this case, the first iteration of (1) is also $W_i^d(t = R_i^0)$. If $W_i^d(R_i^0) \leq D_i - J_i$, then we know that $R_i \leq W_i^d(R_i^0) \leq D_i - J_i < R_i^0$, therefore task $\tau_i$ is schedulable. The smaller the initial value, the smaller the upper bound. However, if $W_i^d(R_i^0) > D_i - J_i$, then nothing can be concluded.

However, if task $\tau_i$ is in fact not schedulable, i.e., $0 < D_i - J_i < R_i$ then any initial value taken from the interval $(0, +\infty)$ will cause the iteration to converge at an upper bound of $R_i$ which is greater than $D_i - J_i$. Since the test is only sufficient and there is no prior knowledge about the schedulability of task $\tau_i$, we can't decide the schedulability of task $\tau_i$. Thereafter exact schedulability tests have to be provided to examine its schedulability. So for the extra computation time of the sufficient test, the greater the initial value, the less the number of iterations which leads to the condition $R_i > D_i - J_i$.

According to above discussion, we can observe that for a sufficient test, the initial value can be arbitrary positive value. If the task is schedulable, an appropriate initial value can help the test succeed fast and save computations; if the initial value is unsuitable, the test will fail and the computation is wasted in which case the exact test must be performed. If the task is not schedulable, then the sufficient is doomed to fail. In this case, selecting a larger initial value to make the test fail earlier can reduce the wasted computation time.

It is clear that for the initial value $R_i^0 < R_i$, the larger it is, the less the number of iterations Eq. (1) takes. When $R_i^0 > R_i$, the number of iterations Eq. (1) takes is uncertain. Figure 1 shows the simulation results of the maximum number of iterations when initial values are selected from $R_i^0 > R_i$ compared with the number of iteration when the initial value is $R_i^0 = C_i + B_i$ for total utilizations varying from 0.65 to 0.95. We randomly generate 1000 task sets each with 100 tasks. The x label is the task number. Different lines in Fig. 1 represent different utilization factors. We can observe that the number of iterations when $R_i^0 > R_i$ is basically less than half of the initial value $R_i^0 = C_i + B_i$.

Based on above observations, we believe that the efficiency of schedulability tests can be improved by searching for better initial values within $[R_{i,exact}^0, D_i - J_i]$, where $R_{i,exact}^0$ is the existing initial values for the exact test. So the key problem is to find better step size in search. In the following, we present the selection of initial values based on the existing ones.

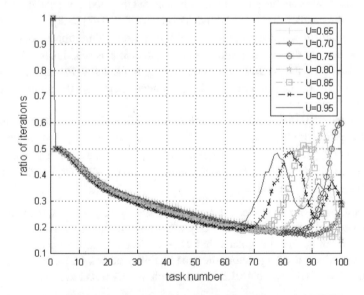

**Fig. 1** The number of iterations when the initial value is bigger than tasks' WCRT

## 3.2 Initial Values for Response Time Analysis

Sjödin and Hansson [15] introduced the following closed form lower bound:

$$R_i^{LB} = \frac{C_i + B_i + \sum\limits_{\forall j \in hp(i)} J_j U_j}{1 - \sum\limits_{\forall j \in hp(i)} U_j}. \tag{4}$$

Based on [16], Davis and Burns [17] extended the response time upper bound with release jitter:

$$R_i^{UB} = \frac{C_i + B_i + \sum\limits_{\forall j \in hp(i)} (J_j U_j + C_j(1 - U_j))}{1 - \sum\limits_{\forall j \in hp(i)} U_j}. \tag{5}$$

It is instinctive to select the initial value from $[R_i^{LB}, R_i^{UB}]$, we can write

$$R_i^{\#1} = \frac{C_i + B_i + \sum\limits_{\forall j \in hp(i)} (J_j U_j)}{1 - \sum\limits_{\forall j \in hp(i)} U_j} + \left( \frac{\sum\limits_{\forall j \in hp(i)} (C_j(1 - U_j))}{1 - \sum\limits_{\forall j \in hp(i)} U_j} \right) \delta_1, \tag{6}$$

where $R_i^{\#1}$ is the proposed initial value and the parameter $\delta_1 \geq 0$ is tunable. Notice that if $\delta_1 = 0$, then there is $R_i^0 = R_i^{LB}$ and if $\delta_0 = 1$, then there is $R_i^0 = R_i^{UB}$. When $\delta_1 > 0$, the test is only sufficient.

The term $1 - \sum_{\forall j \in hp(i)} U_j$ costs a lot of computation and more importantly it varies with $i$. To be simple and to reduce computation, We can simply give the parametric initial value as follows:

$$R_i^{\#1} = (C_i + B_i + \sum\limits_{\forall j \in hp(i)} (J_j U_j + C_j(1 - U_j))) \delta_1. \tag{7}$$

Based on [18], Davis [7] extended their initial value with blocking time and release jitter and derived an improved initial value $(D_i - J_i + C_i + B_i)/2$. They also proved that it can be considered optimal in the sense that the initial value is tight; any increase in this value could, in the general case, result in the schedulability test ceasing to be exact. We extend it with the following parametric initial value for the sufficient test:

$$R_i^{\#2} = (D_i - J_i + C_i + B_i) \delta_2, \tag{8}$$

where $R_i^{\#2}$ is the proposed initial value and the parameter $\delta_2$ is tunable. So if $0 < \delta_2 \leq 0.5$, the test is exact and if $\delta_2 > 0.5$, it is only sufficient.

**Fig. 2** Matlab code of the
RTA algorithm

```
 1  wprev = 0;
 2  w0 = C(i) + B(i);
 3  w = initial_value_sufficient;
 4  while w > wprev && w <= D(i) - J(i)
 5      wprev = w;
 6      w = w0;
 7      for j = 1 : i-1
 8          w = w + ceil((wprev + J(j)) / T(j)) * C(j);
 9      end
10  end
11  if w > D(i)-J(i)
13  ······ %%-- test again with w=initial_value_exact;
14  end
```

The initial value $R_{i-1} - B_{i-1} + B_i + C_i$ [15] is also an effective initial value but it needs to test the task's schedulability in priority order, highest priority first. We extend it as follows:

$$R_i^{\#3} = (R_{i-1} - B_{i-1} + B_i + C_i)\delta_3, \tag{9}$$

where $R_i^{\#3}$ is the proposed initial value and the parameter $\delta_3$ is tunable. If $0 < \delta_3 \leq 1$, the test is exact and if $\delta_3 > 1$, it is only sufficient.

A standard implementation of the fix-point iteration given by (1) is shown in Fig. 2. The Matlab code fragment from line 1 to line 10 computes the response time of the task $\tau_i$.

To test schedulability, we first use the sufficient condition, after it fails, the exact one is used. We use the proposed initial values with parameters' values $\delta_1 > 0$, $\delta_2 > 0.5$, and $\delta_3 > 1$ respectively for the sufficient test. If the calculated response time or response time upper bound of the task $\tau_i$ is less than or equal to $D_i - J_i$, then task $\tau_i$ is schedulable. If it is bigger than $D_i - J_i$, then we use the initial values $R_i^{LB}$ of (3), $(D_i - J_i + C_i + B_i)/2$, or $R_{i-1} - B_{i-1} + B_i + C_i$. instead for the exact test and decide if task $\tau_i$ is schedulable or not.

It is interesting to note that if task $\tau_i$ is in fact not schedulable, since the initial value $R_i^{LB}$ is a lower bound of $R_i$, then if $R_i^{LB} > D_i - J_i$, we can directly conclude that task $\tau_i$ is not schedulable and the algorithm does not need to iterate. Since different delta parameters $(\delta_1, \delta_2, \delta_3)$ represents different initial values $(R_i^{\#1}, R_i^{\#2}, R_i^{\#3})$ that yield different performance for a particular task set, we can test the schedulability of the task set with different delta parameters offline to find most efficient one for online use.

## 3.3  Initial Value for Processor Demand Analysis

As discussed in the end of Sect. 2, another way to perform the schedulability test is
to traverse every $t \in [0, D_i - J_i]$ to find some $t$ satisfying $W_i^d(t) \leq t$. Based on the
observation that $W_i^d(t)$ is piecewise strictly monotonically increasing at a finite set
of values of $t$, only a finite time points should be tested [3]. We give the test
condition in the following equation:

$$C_i + B_i + W_{i-1}^d(t) \leq t, (R_i^{\#4} = t = (D_i - J_i)\delta_4), \qquad (10)$$

where $R_i^{\#4}$ is the initial value or starting test point and the parameter $0 < \delta_4 \leq 1$ is
tunable. Note that within $[0, R_i]$, there is always $W_i^d(t) \leq t$, so $R_i^{\#4}$ is used from
bigger values to small ones in the search.

The implementation of the test condition in (10) is shown in Fig. 3. The test is
performed twice. We first test (10) with $t = (D_i - J_i)\delta_4^1$, if (10) is satisfied, then we
conclude that task $\tau_i$ is schedulable. But if the calculated demand is bigger than $t$,
then we use $t = (D_i - J_i)\delta_4^1\delta_4^2$ instead and test again. Since we can let $\delta_4$ strictly
less than one, if (10) is satisfied in the second test, then we can also conclude that
task $\tau_i$ is schedulable. This procedure can keep going until (10) is satisfied. So we
let the test continue $k$ times ($\delta_4$ may own a series of values, $\delta_4^1, \delta_4^2, \ldots, \delta_4^k$), if all
have failed, then the exact schedulability test is used instead to calculate the
response time and decide if task $\tau_i$ is schedulable or not.

```
1   w0 = C(i) + B(i);
2   w = w0;
3   wprev = (D(i) - J(i))*delta4(1);
4   for j = 1:i-1
5       w = w + ceil((wprev + J(j))/T(j))*C(j);
6   end
7   if w > wprev;
8       wprev = wprev*delta4(2);
9       w = w0;
10      for j = 1:i-1
11          w = w + ceil((wprev + J(j))/T(j))*C(j);
12      end
13      if w > wprev
14          %%-- run exact test with w=initial_value_exact;
15      end
16  end
```

**Fig. 3** Matlab code of TDA algorithm with $R_i^{\#4}$

## 3.4   Implementation

From above discussion, we obtain four initial values with delta parameters to adjust. Enumerating proper parameters for all kinds of task sets is impossible. Since $\delta_k$ is a continuous variable, even for a given task set, the effort to find a best value through testing all possible values of $\delta_k$ will fail anyway. The use of a unified mathematical formulation to calculate a best solution of $\delta_k$ can also be difficult because the performance of a schedulability test depends on the details of tasks' specific parameters. So we mainly search for proper values of $\delta_k$ utilizing discrete-event scheduling simulation.

## 4   Evaluations

In this section, we report on the numerical experiments to give some tips about how to use the proposed initial values and also to compare our results with other known ones.

## 4.1   Stochastic Model

We randomly generated task sets with constrained deadlines. Unbiased utilizations were generated using the UUniFast algorithm [19]. Period $T_i$ was randomly generated in the interval [1, 2500] and sorted in ascending order. The worst-case execution time $C_i$ was computed as $C_i = U_i T_i$. The worst-case blocking time $B_i$ was randomly generated in the interval $[0, C_{i-1}^{\max}]$ and also sorted in ascending order, where $C_{i-1}^{\max}$ is the maximum value of $C_k(1 \leq k \leq i-1)$. Note that $C_{i-1}^{\max}$ is an upper bound of $B_i$. The release jitter $J_i$ was randomly generated within the interval $[0, 0.05 \cdot T_i]$, i.e., less than 0.05 times its period. Deadline $D_i$ was assumed to be equal to its period $D_i = T_i$.

Experiment parameters are the number of tasks $n$ in a task set, the utilization factor $U$ and the proposed parameters $\delta_k(k = 1, 2, 3, 4)$. For fixed parameters, every experiment was replicated 5000 times in order to achieve unbiased statistics. The utilization factor varies from 0.6 to 0.95 (step 0.05) and for every value, the same number of task sets has been generated. The number of tasks $n$ varies from 50 to 400 (step 50). As for the parameter $\delta_4$, we repeated the sufficient schedulability tests [e.g., test (10)] twice, if both have failed, then the exact test was used.

## 4.2 Performance Analysis

The simulation is done using Matlab. It is shown in [20] that response time analysis tests outclass scheduling points tests as far as the analysis time is concerned. So we use the response time analysis technique as the exact test in our experiment, which is illustrated in Fig. 2, i.e., the standard implementation of the RTA algorithm. We compare our algorithm with the initial value $C_i + B_i$, the initial value $(D_i - J_i + C_i + B_i)/2$ which is proven more efficient than others (Table 4 from [7]) and the initial value $R_{i-1} - B_{i-1} + B_i + C_i$. Note that the acceptance ratios are all the same as the exact test because the sufficient tests are substituted by the exact test if they have failed. The computation time ratios are the computation times of schedulability tests to the exact test with initial value $C_i + B_i$ measured by Matlab on a PC. The number of tasks in a task set is 100 for Figs. 4 and 5.

Note that in the experiments, the response times are calculated in the priority order in a task set, and after the sufficient condition fails, first $(D_i - J_i + C_i + B_i)/2$ is used for the exact test until some task is proved to be not schedulable, then $R_{i-1} - B_{i-1} + B_i + C_i$ is used for the exact test thereafter.

Figure 4 shows the performance of schedulability test using initial value $R_i^{\#2}$ with $\delta_2 = 0.5, 0.7, 0.9$ respectively. It is observed that the factor $\delta_2 = 0.5$ performs worst. With the factor $\delta_2 = 0.7$, the schedulability test performs a little better than the factor $\delta_2 = 0.9$ when total utilization factor $U$ is less than or equal to 0.75 and worse when $U > 0.75$. From the result, we can observe that the efficiency of schedulability test can be further improved by choosing more appropriate

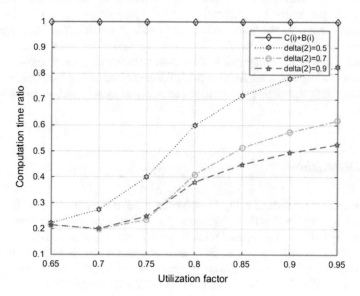

**Fig. 4** The performance of initial value $R_i^{\#2}$ with different $\delta_2$

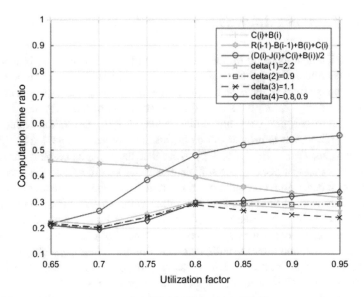

**Fig. 5** The performance of proposed method with varying utilizations

parameters for one particular utilization. The case is similar for $R_i^{\#1}$, $R_i^{\#3}$ and $R_i^{\#4}$, and for the sake of simplicity, we do not illustrate other results here.

Figure 5 shows the computation times of our proposed schedulability tests compared with two exact tests with initial values $(D_i - J_i + C_i + B_i)/2$ and $R_{i-1} - B_{i-1} + B_i + C_i$ respectively. We can observe that with the selected parameters illustrated in the figure, the performances of the four proposed tests are nearly all better than exact ones for utilization factor from 0.65 to 0.95. The improvement is as high as 25 % for $U = 0.8$.

Figure 6 shows the computation time ratio with the number of tasks $n$ varying from 50 to 400 in a task set for $U = 0.85$. We can observe that the fluctuations are very small. This shows that our method can apply to system with different number of tasks.

## 4.3 Discussion

All experiments were done using Matlab on a PC and the computation times were measured by Matlab functions "tic" and "toc". We have tried our best to avoid errors by repeating every experiment 5000 times, avoiding function call overheads, running the schedulability test in different orders and so on. Besides we have also tried different blocking times, release jitters and larger range of task periods. All the results are similar. So we believe that our conclusions are credible.

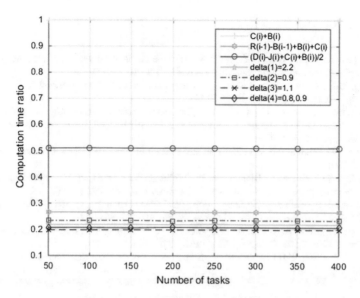

**Fig. 6** The performance of proposed method with varying numbers of tasks

We can further improve the performance of schedulability tests by picking up better parameters. Implementing the experiment on a real hardware platform and improving the adaptability of the methods will be our future work.

## 5 Conclusion

In this paper, we present four parameters used to decrease the time spent for schedulability analysis of fixed priority real-time systems. The proposed tests are based on the observation that more efficient initial values can be found through picking up better parameters which has been supported by experimental results. This proposed approach is to help engineers find better schedulability tests for a particular system offline, which thereafter can be used online. Our proposed techniques have three advantages: the actual less analysis time, the convenience of usage and the adaptability for various task sets. It is observed that with different system utilizations one test performs different with the same parameter. To find best solution for each utilization is needed. Our method can also help find response upper bounds but it is not appropriate if exact response times are required.

# References

1. Yang, M.F., Gu, B., Guo, X.Y., et al.: Aerospace embedded software dependability guarantee technology and application. Sci. Sin. Tech. **45**, 198–203 (2015)
2. Yang, M.F., Wang, L., Gu, B., Zhao, L.: The application of CPS to spacecraft control systems. Aerosp. Control Appl. **38**(5), 8–13 (2012)
3. Lehoczky, J.P., Sha, L., Ding, Y.: The rate-monotonic scheduling algorithm: exact characterization and average case behavior. In: IEEE Real-Time Systems Symposium (RTSS), pp. 166–171 (1989)
4. Joseph, M., Pandya, P.K.: Finding response times in a real-time system. Comput. J. **29**, 390–395 (1986)
5. Audsley, N.C., Burns, A., Richardson, M., Wellings, A.J.: Applying new scheduling theory to static priority pre-emptive scheduling. Softw. Eng. J. **8**, 284–292 (1993)
6. Eisenbrand, F., Rothvoss, T.: Static-priority real-time scheduling: response time computation is NP-hard. In: IEEE Real-Time Systems Symposium (RTSS) (2008)
7. Davis, R.I., Zabos, A., Burns, A.: Efficient exact schedulability tests for fixed priority real-time systems. IEEE Trans. Comput. **57**(9), 1261–1276 (2008)
8. Bini, E., Nguyen, T.H.C., Richard, P., Baruah, S.K.: A response-time bound in fixed-priority scheduling with arbitrary deadlines. IEEE Trans. Comput. **58**(2), 279–286 (2009)
9. Bini, E., Parri, A., Dossena, G.: A quadratic-time response time upper bound with a tightness property. In: IEEE Real-Time Systems Symposium (RTSS) (2015)
10. Fisher, N., Goossens, J., Nguyen, T.H.C., Richard, P.: Parametric polynomial-time algorithms for computing response-time bounds for static-priority tasks with release jitters. In: IEEE International Conference on Embedded and Real-Time Computing Systems and Applications (RTCSA) (2007)
11. Nguyen, T., Richard, P., Bini, E.: Approximation techniques for response-time analysis of static-priority tasks. Real-Time Syst. **43**, 147–176 (2009)
12. Lu, Q., Cheng, A.M.K., Davis, R.I.: Tunable response time upper bound for fixed priority real-time systems. In: IEEE Real-Time and Embedded Technology and Applicaions Symposium (WIP session) (2015)
13. Liu, C.L., Layland, J.W.: Scheduling algorithms for multiprogramming in a real-time environment. J. ACM **20**, 46–61 (1973)
14. Lehoczky, J.P.: Fixed priority scheduling of periodic task sets with arbitrary deadlines. In: IEEE Real-Time Systems Symposium (RTSS) (1990)
15. Sjödin, M., Hansson, H.: Improved response time analysis calculations. In: IEEE Real-Time Systems Symposium (RTSS), pp. 399–408 (1998)
16. Bini, E., Baruah, S.K.: Efficient computation of response time bounds under fixed-priority scheduling. In: International Conference on Real-Time Networks and Systems, pp. 95–104 (2007)
17. Davis, R.I., Burns, A.: Response time upper bounds for fixed priority real-time systems. In: IEEE Real-Time Systems Symposium (RTSS) (2008)
18. Lu, W.C., Lin, K.J., Wei, H.W., Shih, W.K.: Period-dependent initial values for exact schedulability test of rate monotonic systems. In: International Parallel and Distributed Processing Symposium (IPDPS) (2007)
19. Bini, E., Buttazzo, G.C.: Measuring the performanc of schedulability tests. Real-Time Syst. **30**, 129–154 (2005)
20. Min-Allah, N., Khan, S.U., Ghani, N., Li, J., Wang, L.Z., Bouvry, P.: A comparative study of rate monotonic schedulability tests. J. Supercomput. **59**, 1419–1430 (2012)

# A GPU Accelerated Finite Differences Method of the Bioheat Transfer Equation for Ultrasound Thermal Ablation

Georgios Kalantzis, Warner Miller, Wolfgang Tichy
and Suzanne LeBlang

**Abstract** Over the years, high intensity focused ultrasound (FUS) therapy has become a promising therapeutic alternative for non-invasive tumor treatment. The basic idea of FUS therapy is the elevation of the tissue temperature by the application of focused ultrasound beams to focal spot in the tumor. Biothermal modeling is utilized to predict dynamic temperature distributions generated and altered by the therapeutic heating modality, tissue energy storage and dissipation, and blood flow. Implementation of biothermal modeling in the planning, monitoring, control and evaluation of MR guided Focused Ultrasound (MRgFUS) therapies can help to minimize treatment time, maximize efficacy, and ensure the safety of healthy normal tissues, while increasing clinical confidence in MRgFUS treatments. Fast calculations of thermal doses can support in planning, conduction, and monitoring of such treatments. In the current study a GPU-based method in Matlab is proposed, for fast calculations of the temperature and cumulative equivalent minutes at 43° (CEM 43°) based on the bioheat equation. The performance of our proposed method was assessed with three GPUs (GTX 750, GTX 770 and Tesla C2050) for five grid sizes. The maximum speedup was achieved with the Tesla C2050 ($\sim 29$) while GTX 750 demonstrated the lower performance ($\sim 15$).

**Keywords** Thermal therapy · GPU · Focused ultrasound · Thermal dose

G. Kalantzis (✉)
Department of Physics, Florida Atlantic University,
777 Glades Rd, Boca Raton, FL 33431, USA
e-mail: gkalan@gmail.com

W. Miller · W. Tichy
Department of Physics, Florida Atlantic University,
Boca Raton, FL 33431, USA

S. LeBlang
University MRI & Diagnostic Imaging Centers,
Boca Raton, FL 33431, USA

© Springer International Publishing Switzerland 2016
R. Lee (ed.), *Software Engineering, Artificial Intelligence, Networking
and Parallel/Distributed Computing*, Studies in Computational
Intelligence 653, DOI 10.1007/978-3-319-33810-1_4

# 1    Introduction

Radiofrequency (RF) electromagnetic fields or mechanical acoustic waves are used for a wide variety of medical purposes. In many applications, they can be associated with some degree of local tissue heating. The resulting temperature increase depends on the intensity and distribution of the acoustic fields and the thermal properties of the tissue (i.e. thermal conductivity, blood perfusion and heat capacity). For therapeutic applications the objective is to create a substantial temperature increase (to 40–50 °C in hyperthermia [1, 2] or 75–90 °C in thermal ablation [3, 4]).

Minimally invasive thermal ablative therapy as an alternative to conventional surgery in the treatment of solid tumors and other pathologies is increasing in use because of the potential benefits of performing these procedures in an outpatient setting with reduced complications and comorbidity. In ultrasound thermotherapy (USTT), High Intensity Focused Ultrasound (HIFU) is used to achieve focal temperatures of about 90 °C in tissue within a few seconds of heating time. The induced elevated temperature results in the coagulation of the tumor [5] and apoptotic cell death [6]. HIFU surgery has been demonstrated experimentally [7], explained theoretically [8] and practiced clinically [9–11].

HIFU noninvasive cancer therapies facilitates three-dimensional dynamic temperature estimations based on biothermal models. Over the years, the effects of blood flow on heat transfer in living tissue have been studied by many researchers and a large number of bioheat transfer models have been developed on the basis of two main approaches: the continuum approach and the discrete vessel (vascular) approach. In the continuum approach, the thermal impact of all blood vessels models with a single global parameter; and the vascular approach models the impact of each vessel individually [12]. The most widely used continuum model of perfuse tissue was introduced in 1948 by Harry Pennes. The Pennes model [13] was initially developed for predicting heat transfer in the human forearm. Due to the simplicity of the Pennes bioheat model, it was implemented in various biological research works such as for therapeutic hyperthermia for the treatment of cancer [14].

Those biothermal models play crucial role in HIFU treatment because of the desire to optimize the implementation of treatments ahead of time and can be used to identify locations of concern based on safety constraints intended to protect healthy tissues and critical structures, which may be challenging due to their proximity to the target. Biothermal models are also utilized during pretreatment planning to make treatments faster and more effective enabling the balance of healthy tissue constraints and efficient tumor treatment without excessive overdosing. In addition, during the treatment, biotherrmal models can be used in model predictive control to provide treatment feedback and impromptu adjustments for optimal therapy.

Previous studies have suggested numerical methods for accelerated numerical solutions of the bioheat thermal equation (BHTE). Dillenseger and Esneault had proposed an explicit solution of the BHTE in the spatial frequency domain by utilizing the Fourier transformation [15]. Alternatively, parallelized numerical schemes on GPUs have been developed in the past for computationally efficient simulations [16]. In this paper we deal with the numerical simulations of the heat development caused by focused ultrasound on GPUs with Matlab. The main advantages of Matlab are its intuitive higher-level syntax, advanced visualization capabilities and the availability of toolboxes which make easy the implementation of various numerical methods. In the following sections, we discuss the simulation model we employed which consists of a symmetrical heat source and the heat diffusion in the tissue. Finally we report our results for a set of simulations over three different GPUs and conclude with a discussion and future directions.

## 2 Biothermal Distribution of Focused Ultrasound

### 2.1 Bioheat Transfer Equation

Pennes bioheat model is based on four simplifying assumptions:

1. All pre-arteriole and post-venule heat transfer between blood and tissue is neglected.
2. The flow of blood in the small capillaries is assumed to be isotropic. This neglects the effect of blood flow directionality.
3. Larger blood vessels in the vicinity of capillary beds play no role in the energy exchange between tissue and capillary blood. Thus the Pennes model does not consider the local vascular geometry.
4. Blood is assumed to reach the artioles supplying the capillary beds at the body core temperature. It instantaneously exchanges energy and equilibrates with the local tissue temperature.

Based on these assumptions, Pennes modeled blood effect as an isotropic heat source or sink which is proportional to blood flow rate and the difference between the body core temperature and local tissue temperature. The effects of blood perfusion and metabolism on the energy balance within the tissue can be written in this simplified form:

$$\frac{\partial T}{\partial t} = D \nabla^2 T - \frac{T_a - T}{\tau} + Q(x, y, z) + Q_m \tag{1a}$$

$$D = \frac{k}{\rho \cdot C_p} \tag{1b}$$

where T(x, y, z) is the temperature, D is the thermal diffusivity, Q(x, y, z) is the ultrasound heat source and $Q_m$ is the heat generated by metabolic process ($Q_m$ was assumed to be zero in the current study). The perfusion time constant is defined as $\tau = \rho_b \cdot C_v / w_b \cdot C_{vb}$, where $\rho_b$ is the density of blood, $w_b$ is the blood flow rate specified in terms of mass/volume-time, $T_a$ represents the temperature of arterial blood and $C_{vb}$ is the volume specific heat of blood. The volume specific heat and density of blood used in the work described here were 4.2 J cm$^{-3}$ C$^{-1}$ and 1 g mL$^{-1}$, respectively; a moderate blood flow was considered for a perfusion length of 10 mm. Under these conditions the corresponding $\tau$ was 714 s. Finally k is the heat conductivity, $C_p$ is the specific heat of tissue and $\rho$ is the tissue density.

## 2.2  Heat Source Model

Since the motivation of the current study is the performance evaluation of the numerical solution of the bioheat equation, we assumed a simplified thermal source for the ultrasound. It is consisted of an overall intensity term as well as a Gaussian term which is described as:

$$Q(x, y, z) = \beta \cdot e^{-2\left((x-x_0)^2 + (y-y_0)^2 + (z-z_0)^2\right)/w_0^2} \tag{2a}$$

$$\beta = \frac{2\alpha \cdot f \cdot I_0}{C_p \rho} \tag{2b}$$

where $\alpha$ is the tissue absorption coefficient (dB/cm at 1 MHz), f is the ultrasound frequency used for treatment and $I_0$ is the peak intensity of ultrasound at the focal point $(x_0, y_0, z_0)$. Figure 1 illustrates the three dimensional heat distribution of a ultrasound source of 1 MHz. The temperature is in °C and the axis is in mm. Table 1 reports the parameters values used in the current study [17–19].

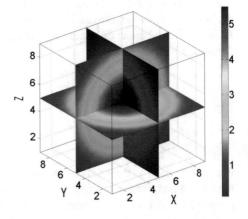

**Fig. 1** 3D heat distribution of ultrasound source with frequency 1 MHz. Axis are scaled in mm and temperature in °C

**Table 1** Literature values for relevant parameters

| Parameter | Value | Units |
|---|---|---|
| k (heat conductivity) | 0.5 | Watss/m/°C |
| $C_p$ (specific heat of tissue) | 3600 | Watts·s/kg/°C |
| P (tissue density) | 1000 | kg/m$^3$ |
| $f_0$ (US frequency) | 1 | MHz |
| α (tissue absorption) | 0.7 | Db/cm@1 Hz |
| $I_0$ (US intensity) | 1500 | Watts/cm$^2$ |
| $w_0$ (beam waist) | 0.5 | cm |

## 2.3 CEM 43 °C Thermal dose Isoeffect Model

The extent of thermal damage to tissue depends on tissue sensitivity, temperature and exposure time. In vitro studies showed that the rate of cell death is exponential with respect to temperature over a limited temperature range. The effect of a given temperature profile on tissue can be determined using Sapareto and Dewey's function [20]. This function calculates the cumulative equivalent minutes at 43 °C (CEM 43 °C) as a model to estimate the thermal isoeffect dose of the actual thermal exposure. Using this method any time-temperature history is converted to an equivalent number of minutes of heating at 43 °C, using the following formula:

$$CEM\,43\,°C = \int_0^t R^{T_{ref}-T(t)}\,dt \qquad (4a)$$

$$R = \begin{cases} 0.5, T \geq T_{ref} \\ 0.25, T < T_{ref} \\ T_{ref} = 43\,°C \end{cases} \qquad (4b)$$

Figure 2 shows the CEM 3 °C as a function of temperature for a heat exposure of 30 min. The model illustrates the exponential increase of CEM 43 °C and consequently the normal tissue damage with increasing temperature.

**Fig. 2** Cumulative equivalent minutes at 43 °C as a function of the temperature for a heat exposure of 30 min

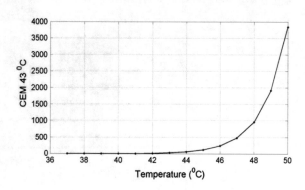

## 2.4 GPU-Based Parallelization

We solve the dynamic bioheat equation using an explicit forward Euler time integration scheme; therefore, we can update all temperature values in the simulation grid simultaneously. First, we determine the spatial distribution of the heat source in three dimensions. Then, we invoke the bioheat kernel for the whole simulation domain, and we use two copies in a ping-pong fashion as read- or write-target, alternately. To respect the boundary conditions, we update only non-boundary voxels in the simulation domain. Since in Matlab there is no direct access to GPU threads instead the parallelization of the algebraic operations are done internally by utilizing the high performance CUBLAS library [21], we followed an indexing approach as shown in Fig. 3. Scalar arrays with the indices were created and they were used for the time integration of the discrete Laplacian operator as follows:

$$\nabla^2 T^{n+1}(x1, y1, z1) = \frac{1}{\Delta x^2} (T^n(x2, y1, z1) + T^n(x3, y1, z1)$$
$$+ T^n(x1, y2, z1) + T^n(x1, y3, z1) + T^n(x1, y1, z2)$$
$$+ T^n(x1, y1, z3) - 6T^n(x1, y1, z1)) \tag{5a}$$

*where*

$$x1 = y1 = z1 = 2 : n+1 \tag{5b}$$

$$x2 = y2 = z2 = 1 : n \tag{5c}$$

$$x3 = y3 = z3 = 3 : n+2 \tag{5d}$$

**Fig. 3** Three dimensional stencils for the numerical implementation of the forward Euler method for the bioheat equation

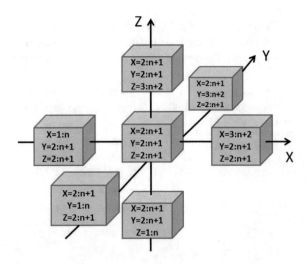

The motive of using scalar arrays with the indices is the reduction of the required *for* loops, which in Matlab are computationally slow. In that way the implementation of the forward Euler in three dimensions, required only one *for* loop just for the time integration. Furthermore, the parallelization of the code on the GPU was almost identical as the serial one, since Matlab utilizes all the available resources of the GPU internally with the proposed indexing approach.

## 3 Results for Homogeneous Environment

### 3.1 Materials and Elements

In this section we introduce our computation environment and report our results of our method. The serial code was implemented in MATLAB and was launched on a desktop with a quad core Intel Xeon X5550 at 2.67 GHz with 6 GB of RAM. For the parallelization on the GPU the parallel computing toolbox was employed and the code was launched on (i) a GTX 750 with Kepler architecture, with 4 SM utilizing in total 512 CUDA cores and 1 GB of GDRR5, (ii) a GTX 770 with Kepler architecture, 1,536 CUDA cores grouped into 8 SMs operating at 1046 MHz and 2 GB of GDRR5 global memory and (iii) a Tesla C2050 with Fermi architecture, 448 CUDA cores arranged into 14 SMs operating at 575 MHz and 3 GB of GDRR5. The performance comparison was established on the speedup factors as described by Eq. 6.

$$speedup = \frac{T_{CPU}}{T_{GPU}} \tag{6}$$

### 3.2 Heat Distribution at the Focal Spot

Figure 4 demonstrates the on-time of the heat source with total duration of 10 s. The temperature at the focal spot rises from the initial temperature of 37 to 77.5 °C

**Fig. 4** Simulated heat source with duration 10 s

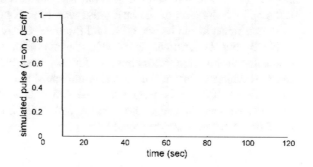

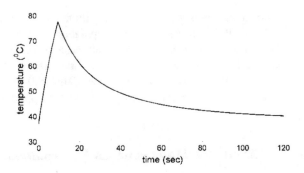

**Fig. 5** Temperature at the location of the focal spot with duration 10 s

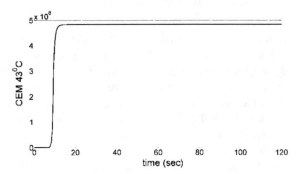

**Fig. 6** CEM 43 °C for a heat pulse of duration 10 s

in 10 s and then decrease approximately exponentially to the baseline temperature (Fig. 5). In Fig. 6 for illustration purpose the CEM 43 °C is plotted versus time. We notice a rapid increase of the CEM 43 °C at time t $\sim$ 7 s where the temperature is about 70 °C.

## 3.3 Evaluation of the Algorithm Performance

The performance of the GPU code was evaluated for all three GPUs with five different grid sizes: 125, 1000, 3375, 8000 and $15.6 \times 10^3$ cm$^3$. The resolution of the 3D grid was 0.1 cm and the time bin was 0.6 s. The total simulation time was 120 s and the duration of the heat pulse was 10 s. Figures 7, 8 and 9 reports the execution times in sec for the CPU and the three GPUs (Fig. 10).

Finally Fig. 11 illustrates the speedup factors for the set of our simulations. We notice that the speedup factors increase for all three GPUs as the grid size becomes larger, reaching a plateau with a maximum speedup of $\sim 29$ for the Tesla C2050 when $15.6 \times 10^6$ voxels were simulated ($25 \times 25 \times 25$ cm$^3$ grid size). Similar performance was attained with the GTX 770 while GTX 750 shows the smaller speedup factors as it was expected.

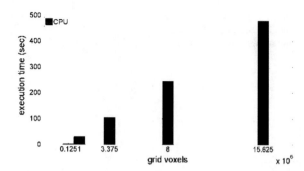

**Fig. 7** Execution times (s) for the CPU as a function of the number of grid voxels

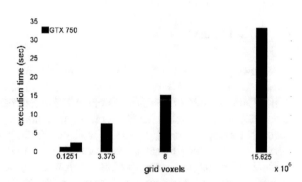

**Fig. 8** Execution times (s) for the GTX 750 as a function of the number of grid voxels

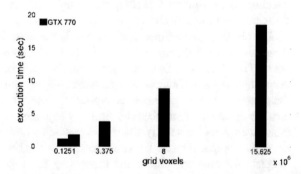

**Fig. 9** Execution times (s) for the GTX 770 as a function of the number of grid voxels

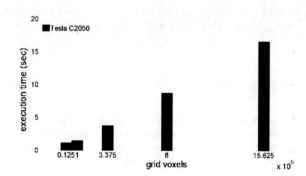

**Fig. 10** Execution times (s) for the Tesla C2050 as a function of the number of grid voxels

**Fig. 11** Speedup factors for
the 3 GPUs as a function of
the number of grid voxels

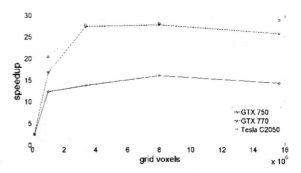

## 4 Conclusions

A GPU-based solution of the bioheat equation for FUS therapy in Matlab was
presented. Three GPUs were evaluated in the current study for five grid sizes.
A maximum speedup of $\sim 29$ was achieved for the Tesla C2050 for the larger grid
size ($15.6 \times 10^6$ voxels). Similar performance attained with the GTX 770 while
speedup factors reduced by $\sim 48$ % with the GTX 750. Future directions of the
current work include thermal dose calculations based on MR thermometry. That
method relies on the water proton resonance frequency (PRF). This approach
provide continuous temperature mapping inside the human body as well as target
tracking information by exploiting the complex nature of the MR-signal and phase
shifts between dynamically acquired phase images and reference data sets.

The abiding goal of the current study is the completeness of a computational
simulator for FUS therapy which could be used for educational and research purposes. Therefore, a realistic heat source is essential for more realistic modeling of
beam intensity distributions and consequently of optimization of treatment. The
focusing of ultrasound waves is typically achieved by a phased system which
consists of several transducer elements. By adjusting the phase shift values of each
transducer, a high intensity focus at a predefined location can be achieved. The
nearfield component can be simulated by evaluating the Rayleigh-Sommerfeld
integrals for each transducer and computing the superposition of them. For the
farfield consideration must be given to the tissue inhomogeneities. Therefore,
corrections should be applied on the pressure fields obtained in the spatial domain
regarding the absorption and speed of sound.

## References

1. Dewhirst, M.W., Vujaskovic, Z., Jones, E., Thrall, D.: Re-setting the biologic rationale for
   thermal therapy. Int. J. Hyperth. **21**, 779–790 (2005)
2. Horsman, M.R., Overgaard, J.: Hyperthermia: a potent enhancer of radiotherapy. Clin. Oncol.
   (R Coll Radiol) **19**, 418–426 (2007)

3. Rempp, H., Hoffmann, R., Roland, J., Buck, A., Kickhefel, A., Claussen, C.D., Pereira, P.L., Schick, F., Clasen, S.: Threshold-based prediction of the coagulation zone in sequential temperature mapping in MR-guided radiofrequency ablation of liver tumours. Eur. Radiol. **22**, 1091–1100 (2012)

4. Malcolm, A.L., ter Haar, G.R.: Ablation of tissue volumes using high intensity focused ultrasound. Ultrasound Med. Biol. **22**, 659–669 (1996)

5. Overgaard, J.: Effect of hyperthermia on malignant cells in vivo. A review and a hypothesis. Cancer **39**, 2637–2646 (1977)

6. Vykhodtseva, N., McDannold, N., Martin, H., Bronson, R.T., Hynynen, K.: Apoptosis in ultrasound-produced threshold lesions in the rabbit brain. Ultrasound Med. Biol. **27**, 111–117 (2001)

7. Fry, F.J., Johnson, L.K.: Tumor irradiation with intense ultrasound. Ultrasound Med. Biol. **4**, 337–341 (1978)

8. Robinson, T.C., Lele, P.P.: An analysis of lesion development in the brain and in plastics by high-intensity focused ultrasound at low-megahertz frequencies. J. Acoust. Soc. Am. **51**, 1333–1351 (1972)

9. Hynynen, K., Pomeroy, O., Smith, D.N., Huber, P.E., McDannold, N.J., Kettenbach, J., Baum, J., Singer, S., Jolesz, F.A.: MR imaging-guided focused ultrasound surgery of fibroadenomas in the breast: a feasibility study. Radiology **219**, 176–185 (2001)

10. Sanghvi, N.T., Foster, R.S., Bihrle, R., Casey, R., Uchida, T., Phillips, M.H., Syrus, J., Zaitsev, A.V., Marich, K.W., Fry, F.J.: Noninvasive surgery of prostate tissue by high intensity focused ultrasound: an updated report. Eur. J. Ultrasound **9**, 19–29 (1999)

11. Gelet, A., Chapelon, J.Y., Bouvier, R., Rouviere, O., Lasne, Y., Lyonnet, D., Dubernard, J.M.: Transrectal high-intensity focused ultrasound: minimally invasive therapy of localized prostate cancer. J. Endourol. **14**, 519–528 (2000)

12. Raaymakers, B.W., Kotte, A.N.T.J., Lagendijk, J.J.: "Discrete vasculature (DIVA) model simulating the thermal impact of individual blood vessels for in vivo heat transfer". In: Minkowycz, W.J. (ed.) Advances in Numerical Heat Transfer, vol. 3, pp. 121–148. CRC Press, Boca Raton, USA (2009)

13. Pennes, H.H.: Analysis of tissue and arterial blood temperatures in the resting human forearm. J. Appl. Physiol. **1**, 93–122 (1948)

14. Minkowycz, W.J., Sparrow, E.M., Abraham, J.P.: Advances in Numerical Heat Transfer, vol. 3. CRC Press, Boca Raton, USA (2009)

15. Dillenseger, J.L., Esneault, S.: Fast FFT-based bioheat transfer equation computation. Comput. Biol. Med. **40**, 119–123 (2010)

16. Georgii, J., von Dresky, C., Meier, S., Demedts, D., Schumann, C.: Focused ultrasound-efficient GPU simulation methods for therapy planning. In: 8th Workshop on Virtual Reality Interactions and Physical Simulations, pp. 119–128 (2011)

17. Canney, M.S., Bailey, M.R., Crum, L.A., Khokhlova, V.A., Sapozhnikov, O.A.: Acoustic characterization of high intensity focused ultrasound fields: a combined measurement and modeling approach. J. Acoust. Soc. Am. **124**, 2406–2420 (2008)

18. Curra, F.P., Mourad, P.D., Khokhlova, V.A., Cleveland, R.O., Crum, L.A.: Numerical simulations of heating patterns and tissue temperature response due to high-intensity focused ultrasound. IEEE Trans. Ultrason. Ferroelectr. Freq. Control **47**, 1077–1089 (2000)

19. Gutierrez, G.: Study of the bioheat equation with a spherical heat source for local magnetic hyperthermia. Mec. Comput. 3562–3572 (2007)

20. Sapareto, S.A., Dewey, W.C.: Thermal dose determination in cancer therapy. Int. J. Radiat. Oncol. Biol. Phys. **10**, 787–800 (1984)

21. Liu, X., Cheng, L., Zhou, Q.: Research and comparison of CUDA GPU programming in Matlab and Mathematica. In: Proceedings of 2013 Chinese Intelligent Automation Conference, pp. 251–257 (2013)

# Justifying the Transition from Trustworthiness to Resiliency via Generation of Safety Cases

Chung-Ling Lin, Wuwei Shen and Steven Drager

**Abstract** Safety analysis plays an important role for developing cyber-physical systems since many of them are also safety critical systems. The failure of cyber-physical systems can have some serious consequences. With the latest development in formal methods, many systems have been converted to a formal model to ensure that all safety requirements have been met. In this case, the systems are called trusted. However, many failures are caused by the missing identification of some properties during the early phase of software development. Thus, a safety case has been widely used as an argument structure to represent how a system has been developed to satisfy safety requirements, and is an important means of communication between various stakeholders in a system. In this paper, we present a novel approach to show how an argument structure can be automatically built via safety case patterns and metamodels underlying a development process. We notice that a transition from trustworthiness to resiliency for many cyber-physical systems is made by separating a fault model from a nominal (non-failure) model in Simulink due to some design considerations such as reduction of a test case generation and the complexity of code. Thus, we take the translation of a nominal model into a fault model into account and employ the model-driven architecture and safety case pattern together to illustrate how a safety case is generated for an argument of the correct transition of a cyber-physical system in Simulink. Last, we discuss how an argument structure of a safety case can be affected by system evolution.

**Keywords** Cyber-physical system · Trustworthiness · Resiliency · Model transformation · Safety-critical systems · Simulink

C.-L. Lin (✉) · W. Shen
Department of Computer Science, Western Michigan University,
Kalamazoo, MI 49008, USA
e-mail: chung-ling.lin@wmich.edu

W. Shen
e-mail: wuwei.shen@wmich.edu

S. Drager
Air Force Research Laboratory, Rome, NY, USA
e-mail: steven.drager@us.af.mil

© Springer International Publishing Switzerland 2016
R. Lee (ed.), *Software Engineering, Artificial Intelligence, Networking
and Parallel/Distributed Computing*, Studies in Computational
Intelligence 653, DOI 10.1007/978-3-319-33810-1_5

# 1 Introduction

Model-based safety analysis plays an important role in developing a cyber-physical system. Many cyber-physical systems are also safety critical in that failure of these systems can have some serious consequences. Thus, in modern software development for a cyber-physical system, the cost to ensure that a system can achieve desired quality-related attributes, which are also called software assurance, accounts for 30–50 % of the total project cost. Despite of the high cost, the current approaches to software assurance, mainly testing and inspection, are regarded as inadequate. Furthermore, testing cannot yield the software assurance for many kinds of errors [1].

For instance, in September 2006, a Royal Air Force (RAF) Hawker Siddeley Nimrod suffered an in-flight fire and subsequently crashed in Afghanistan. Later, the investigators found that some fuel appeared to have leaked into the bomb bay which caught fire and resulted in the final crash. Obviously, both the physical part such as the bomb bay and its sensor and the software system did not cooperate with each other. While the RAF Nimrod had been extensively tested before its service, the crash still happened. Finally, the investigators detected that the critical catastrophic fire hazard had not been appropriately assessed since it was one of those hazards left "Open" and "Unclassified" [2]. Thus, as suggested in [1], to develop a successful software system, developers should succeed both in quality control (preventative methods) and in assuring that it is there (evaluative methods). Only some preventative methods, such as testing and inspection, are not enough to ensure quality control of a system. Developers must employ some evaluative methods to assure that quality control does exist. Furthermore, the evaluative methods should be carried out at the early stage of a software development life cycle (SDLC) to ensure all quality control requirements have been correctly retrieved from the beginning. To support evaluative methods, some notations such as an assessment model should be constructed. In this sense, a goal structuring notation (GSN) as an argument structure has been proposed by Kelly [3] and thus extensively used.

On the other hand, with the increasing complexity of applications for software intensive systems, developers employ Model-Driven Architecture (MDA) to aid in software development. The advantage of MDA is the application of some automation to create, transform, and manage various artifacts produced during SDLC. For a cyber-physical system, developers use MDA to support various activities such as design, simulation, verification, testing, and code generation. For instance, some developers proposed to incorporate the safety analysis activities into the traditional development activities via MDA [4]. In this case, trustworthiness can be achieved via testing and verification methods. To support the transition from trustworthiness to resiliency, Joshi et al. found that system analysis requires knowledge of the various faults that can occur and the different ways in which a system component can malfunction [4]. Thus, the best strategy to design a cyber-physical system is to start with a nominal model, which is proven to be non-failing, and then augment some fault behavior to the nominal model in an extended fault model. Note that some activities such as test-case generation and formal analysis of nominal models becomes complicated and hard on an extended fault model, since it includes more fault behaviors. Thus, the separation of a nominal model and its extended fault model leverages the capability in system understanding, maintenance, creation, and evolution. In reality, a translation from a nominal model to a fault model is a viable means to support the trustworthiness as well as the resilience of a system.

It is challenging to argue that a cyber-physical system is both trustworthy and resilient. Any manual effort can be notoriously difficult and error-prone, thus, some automation of safety cases is an important part of safety analysis for a cyber-physical system. In this paper, we use safety case patterns to illustrate how an argument structure via a safety case can be produced for the translation of a nominal model to a fault model in a cyber-physical system. In particularly, we consider both nominal models and their fault models represented in Simulink. Furthermore, we take the translation from a nominal model to its extended fault model into account and demonstrate how a safety case is generated as an argument structure to show why the translation is correct. Based on a generation of a safety case, we discuss how artifact modification affects the argument structure of a safety case and finally makes an impact on safety analysis.

In this paper we make the following contributions.

- The automatic generation of safety cases to provide an argument structure to support the transition from trustworthiness to resiliency for a cyber-physical system;
- The establishment of a Simulink meta-model which aids in the generation of a safety case;
- Discussion of the impact of artifact modification on the argument structure of a safety case and finally on safety analysis.

This paper is organized as follows. Section 2 introduces the safety case patterns. Section 3 introduces a Simulink meta-model and some safety case patterns to support the resiliency of a cyber-physical system. Section 4 illustrates the implementation of generation of a safety case via a safety case pattern. We discuss how a generated safety case helps the evaluation and system evolution in Sect. 5. We present some related work and draw conclusions in Sect. 6.

## 2 Introduction to Safety Case Patterns

A safety case pattern is abstracted from the details of a particular argument in the context of an argument structure which can be applied in various situations. The application of a safety case pattern leverages the capability of software development for a cyber-physical system in that a safety case pattern increases re usability and reduces the cost to build an argument structure in a new context. Consequently, a safety case pattern takes advantage of a prior successful argument structure and thus avoids the errors which can occur in a new development context.

A safety case pattern is defined based on the Goal Structuring Notation (GSN). Unlike the standard GSN, a safety case pattern allows roles to occur in a node of a safety case. Figure 1 shows a simple example of a safety case pattern. In this case, it consists of two claim nodes or simply claims, denoted as two rectangles. In the top claim, also called a goal, it has two variables enclosed by "{" and "}". In general, an expression in Object Constraint Language (OCL) [5] can be written inside a pair of "{" and "}". As a special case of an OCL expression, a variable can be written and in this case the variable is called a role. There are two types of roles which can be written within a pair of "{" and "}", we use $A to denote a string role and A is a variable role. So, in Fig. 1, the top claim has the variable role X and the string role $A. A variable role can be divided into two groups, a bound variable role and a free variable role. A variable role is bound if the node, where the variable role occurs, has a precedent node or line where the variable role occurs right after the keyword *foreach*. Otherwise, it is called a free variable role, or just a variable role. For instance, in Fig. 1, the variable X and Z are variable roles since they are not bound by any *foreach* expression. However, the variable role y is a bound variable role since it is bound by the *foreach* expression attached to the link connecting the second rectangle which includes y. During an instantiation of the safety case pattern, y denotes each instance of variable role Z associated with the variable role X.

To instantiate a safety case pattern, we need a weaving table and a domain model. A domain model denotes the meta-information required to develop a software system and is denoted by a UML class diagram. A domain model usually models the activities in SDLC, the important structures of artifacts as well as traceability information among artifacts. A weaving table maps (free) variable roles to classes where a safety case should be generated. For a bound variable role, there is no mapping relationship defined in a weaving table. Furthermore, since a string role is replaced by a string value during the instantiation, there is no need to specify a mapping relationship in a weaving table either. A safety case pattern is instantiable if a weaving table specifies mapping relationship for all the free variable roles. Next, we illustrate how a safety case is instantiated/generated via an example shown in Fig. 2. Figure 2i is the domain model which we base on the generation of a safety case. The mapping relationship at the top of Fig. 2ii is assumed to be

**Fig. 1** Safety case pattern

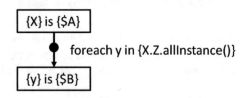

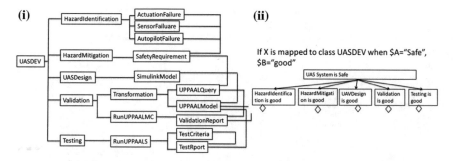

**Fig. 2** A safety case and its domain model. **i** A domain model. **ii** A mapping relationship and a safety case generated

retrieved from a weaving table. Namely, the variable role X is mapped to class UASDEV while the string roles $A and $B are string values "Safe" and "good". Based on the information, we can generate a safety case shown at the bottom of Fig. 2ii.

## 3 Simulink Metamodel and Safety Case Patterns

In this section, we discuss the safety case patterns derived from the failure modes introduced in [4]. The first failure mode is called *Binary_Stuck_at* failure mode for valves. The *Binary_Stuck_at* failure mode for a valve occurs when the valve is stuck in either an open or closed position. According to [4], since the *Binary_Stuck_at* failure mode for a valve needs to access the inputs of the original valve, which is the valve in a nominal model, we need a wrapper around the original valve in a fault model to access the original input. Furthermore, in order to activate the fault mode, each valve has two additional control inputs: Stuck_Flag and Stuck_Val. So, this is illustrated in Fig. 3i. Next, Fig. 3ii shows the class diagram from the Simulink Meta-model. First the class *BinaryStuckAtWrapper* denotes the wrapper. The association between *BinaryStuckAtWrapper* and *Valve* shows that the former is generated based on the latter. The connections between two additional control inputs, modeled by the class *ControlInput,* and the wrapper are denoted by the two associations respectively between the classes *BinaryStuckAt-Wrapper* and *ControlInput.*

Figure 4i shows the safety case pattern for the *Binary_Stuck_at* failure mode for a valve. To support the top claim in the safety case pattern, two sub-claims have been used based on the types of input. The first sub-claim is about the data input while the second sub-claim targets the control input. Following the first sub-claim, a strategy is generated for each nominal data input. Namely, for each nominal data input, the safety case pattern ensures that a corresponding data input is generated in a new fault model. Thus, another sub-claim is produced to argue that a data input is

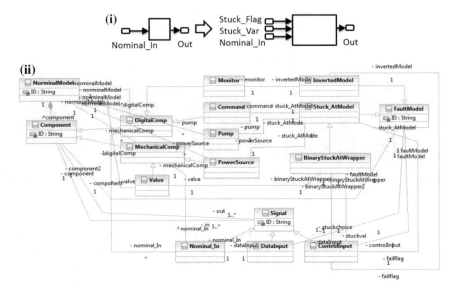

**Fig. 3** Binary_Stuck_At failure mode and Simulink Meta-model. **i** Binary_Struck_At wrapper. **ii** Simulink metamodel

correctly generated in a fault model. Here, we use a context, denoted by an oval connecting to the sub-claim, to show the meaning of correct generation. At the same time, the second sub-claim requires that two control inputs, which are a stuck flag and stuck value respectively, be generated in a fault model.

Likewise, Fig. 4ii shows the safety case pattern for the *Inverted* failure mode and *Stuck_at* failure mode for the digital components and other types of mechanical components such as Pump and Power Source. Last, we introduce the safety case pattern for the generation of a fault model from a nominal model. A fault model

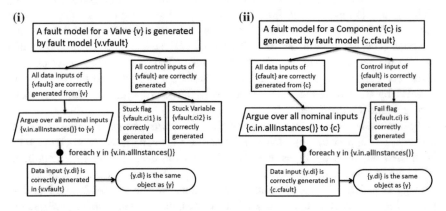

**Fig. 4** Safety case patterns of the failure modes. **i** The Binary_Struck_at pattern. **ii** The inverted and Struck_at failure pattern

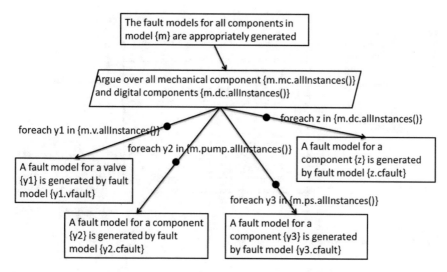

**Fig. 5** Overall safety case pattern

should consider the *Binary_Stuck_at* failure mode for all valves and the *Inverted* failure mode and *Stuck_At* failure mode for the digital components and other types of mechanical components in a nominal model. Therefore, a safety case pattern for the generation of a (correct) fault model is shown in Fig. 5.

## 4    Generation of Safety Case and Wheel Brake System

We developed a framework whose structure is illustrated in Fig. 6. The framework consists of two steps. The first step is to take an input, a safety pattern, and the metamodel mapped by a weaving table given in the pattern. The output of the first step is an ATL program, which is written in a model transformation language called ATL Transformation Language [6]. In the second step, the ATL program takes artifacts from a specific project as input and produces a safety case.

To generate a safety case for a real application, we consider the wheel brake system (WBS) which was initially described in ARP 4761- Appendix L [7] and developed by a team from University of Minnesota and Rockwell Collins Inc. The nominal model of the WBS system, denoted as the WBS nominal model, shown in Fig. 7i, consists of one digital control unit, the BSCU, and two hydraulic pressure lines, Normal (pressured by the Green Pump) and Alternate (pressured by the Blue Pump and the Accumulator Pump) line. The WBS system takes inputs such as PedalPos1, AutoBrake, DecRate, AC_Speed, and Skid. All these inputs are for-warded to the BCSU for computing the brake command. The outputs of the system include Normal_Pressure, Alternative_Pressure, and System_Mode.

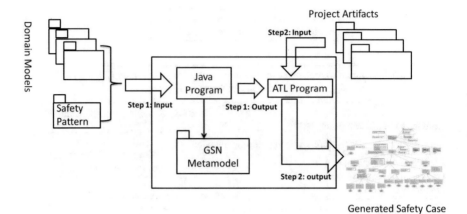

**Fig. 6** An overall structure of the framework

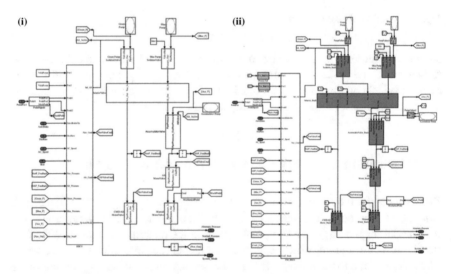

**Fig. 7** Wheel brake system. **i** A nominal wheel brake system in simulink. **ii** Wheel brake system with a fault model

The WBS with a fault model is shown in Fig. 7ii. As mentioned before, the *Binary_Stuck_at* failure mode for each valve has two additional inputs, Stuck_Flag and Stuck_Signal. The other failure modes, such as the Inverted Failure mode or Stuck_At Failure mode for digital components, and other types of mechanical components, have one additional input, Fail_Flag, which activates a fault. Each shaded component in Fig. 7ii represents a mechanical component extended with a failure mode. In particular, the BSCU is composed of two command units and two monitor units, for a total of four additional inputs: Mon1_Fail, Mon2_Fail, Cmd1_Fail, Cmd2_Fail are added to the FM_BSCU in Fig. 7ii.

**Table 1** Weaving table for overall safety case pattern

| Safety case pattern role | Simulink metamodel class |
| --- | --- |
| m | NominalModel |
| mc | MechanicalComp |
| dc | DigitalComp |
| V | Valve |
| pump | Pump |
| PS | PowerSource |
| vfault | FaultModel |
| cfault | FaultModel |

The first step of the framework is to take as an input the safety case patterns with weaving tables and the Simulink Meta-model mapped by the weaving tables to generate an ATL program. In this case study, three safety case patterns are used as inputs, and they are the overall safety case pattern shown in Fig. 5, the *Binary_Stuck_at* pattern shown in Fig. 4i, and the *Inverted* and *Stuck_at* Failure Pattern shown in Fig. 4ii respectively. The Simulink Meta-model is shown in Fig. 3ii. Due to space constraints, we only show a weaving table applied by the overall safety case pattern in Table 1 and skip the weaving tables for the other two patterns. The weaving table specifies how each variable role is mapped to the Simulink Meta-model. For example, the role *m* is mapped to the class *NominalModel* while the role *mc* is mapped to the class in the class *MechanicalComponent* in the Simulink Metamodel. The input meta-model is edited by Rational Software Architect [8] and we output the model into the UML model format. The safety patterns are edited by the graphical GSN editor tool [9] and we output the patterns into the XML format. Our Java program employs the eclipse UML2 [10] to read the UML model and XML DOM parser [11] to parse XML files to retrieve the GSN nodes from safety case patterns to finally produce an ATL program. Our Java program starts with retrieving the first node (i.e. the root node) in a safety case pattern and traverses the rest nodes in the pattern via the GSN meta-model. Specifically, the program executes recursively until all GSN nodes in a safety case pattern are parsed. For each node parsed by the Java program, a corresponding ATL rule is generated. And finally, all ATL rules are put together yield a final ATL program.

A generated ATL program specifies how target model elements, which are a safety case, are produced from source model elements, which are Simulink nominal model as well as its fault model. An ATL program is composed of ATL rules. In our approach, two different types of ATL rules are used in the ATL program: Matched rule and Called rule. For a root goal node of the safety case pattern used by the project, a matched rule in ATL is produced by the Java transformation as output. For the other nodes in the safety case pattern, a called rule is produced for each node by the Java transformation. The main difference between a matched rule and a called rule is that a matched rule needs to explicitly specify the type of the source model elements that will be matched by this rule in its from section, while the matched rule

**Fig. 8** An example of ATL program

```
1 module testmodule;
2 create OUT : GSN from IN : SimulinkModel;
3
4 rule R1 {
5   from
6   s : SimulinkModel!"Simulink::NominalModel"
7   to
8   g1 : GSN!GSN_Goal (
9   )
10
11 do {
12   g1.ID <- 'G1'
13   g1.Content <- '';
14   g1.Content <- g1.Content + 'The fault models for
     all components in model ';
15   g1.Content <- g1.Content + s.ID;
16   g1.Content <- g1.Content + ' are appropriately
     generated ';
17   g1.strategy <- thisModule.called01(s);
18   }
19 }

20 rule called01(p0 : SimulinkModel!NorminalModel) {
     ...
21 }
```

is executed and generates the target model elements only when a matched source model element is found. A called rule in ATL is similar to a method in a programming language that enables it to explicitly generate target model elements in imperative code. However, unlike a matched rule, a called rule must be invoked from an ATL imperative block, which is composed of a sequence of imperative statements similar to statements in a programming language. In the ATL transformation, a matched rule is considered as the starting point for generating a root node for a safety case. The use of called rules in the ATL program allows the invocation from the matched rule so the rest of nodes in the safety case can be produced.

Figure 8 is a partial ATL program generated by the Java transformation as illustration of how the root node of the overall safety case pattern is generated in ATL. First, we declare the two metamodels via keywords *out* and *in* to represent the source and target metamodels in this transformation at line 2 in Fig. 8. Namely, line 2 specifies a target Metamodel via keyword *out*, i.e. the GSN Metamodel shown in Fig. 9, and a source Meta-model via keyword *in*, i.e. the SimulinkModel Metamodel partially shown in Fig. 3ii. Starting from line 4 of Fig. 8, a transformation rule called *R1* is given to generate a root claim/goal node for a safety case according to the overall safety case pattern shown in Fig. 5. Specifically, for every instance of class *NominalModel* in the *Simulink* package, the rule *R1* is produced to generate a root of a safety case. Thus, the *from* section at line 6 declares a variable

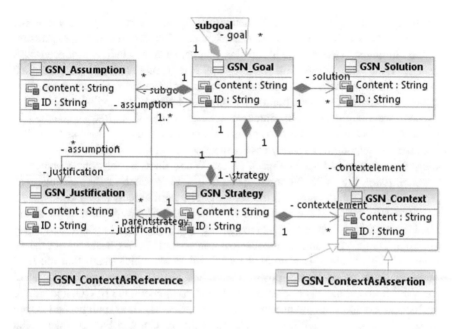

**Fig. 9** GSN Meta-model

*s* whose type is the *NominalModel* class in the *SimulinkModel* Metamodel. Namely, the variable *s* denotes an instance of the class *NominalModel* when the rule is called. The *to* section at line 8 declares a root claim/goal node called the *g1*. Since the type of the *g1* variable is the *GSN_Goal* class in the GSN Meta-model, all attributes of the class *GSN_Goal* should be appropriately assigned to a value according to the root claim of the safety case pattern. In the other word, each goal has two attributes *ID* and *Content* according to Fig. 9. The assignments for both attributes are specify in the *do* section of the rule *R1* from line 11 to line 18. In an ATL program, each attribute is assigned via the binding statement consisting of a feature name, which occurs on the left hand side of the statement, and an expression, which occurs on the right hand side of the statement. For instance, in Fig. 9, a string value "G1", as an expression, is assigned to the *ID* attribute as a feature name. The binding statements from line 13 to 16 produce a string value for a root claim/goal node in the generated safety case according the root claim in the safety case pattern shown in Fig. 5. Note that the root claim of the safety case pattern is "The fault models for all components in model {m} are appropriately generated". Thus, the string the root node of the pattern is divided into three parts separated by the *m* role. The binding assignments at line 14 and 16 produce the sub-string values before and after *m* respectively while the binding assignment at line 15 ensures that a sub-string value of the instance of the mapped class by the *m* role is generated. All sub-string values are combined together via string operation "+" in ATL. Note that in the safety case pattern in Fig. 5, the root claim node is supported by a strategy. Therefore, the binding statement at line 17 sets up a link between the root node and the strategy node based on the GSN metamodel shown in Fig. 9. Specially, the role name *strategy* is used as a feature name of this binding statement. The expression for the binding statement is the invocation of a called rule, which is produced for a strategy node and named *called01()*. The called rule *called01* is generated according to the strategy in the pattern shown in Fig. 5 in the same manner starting at line 20. Due to space constraints, we skip the details of the called rule.

Next, we outline how our framework takes the WBS artifacts as inputs to generate a safety case via the ATL program presented in the previous section. The first claim node/goal in the safety case is the instance of the top claim of the overall safety case pattern shown in Fig. 5. In the WBS, there is only one instance of *NominalModel*, the WBS Nominal. Therefore, one instance of the root claim is generated in the safety case with the content "The fault models for all components in model WBS Nominal are appropriately generated". The root claim of the safety case is supported by a strategy node instance which argues that the fault models for all digital components and all mechanical components in WBS are appropriately generated. This strategy is supported by the sub-goals which claim that the fault model for each valve, pump, power source and digital component is generated. The WBS nominal model consists of one digital component, the BSCU, and twelve mechanical components including two power sources, three pumps and seven valves. For each valve in WBS, an instance of the top claim in Fig. 4i is generated in the safety case. For the rest of components in WBS, the instances of the top claim in Fig. 4ii are generated. Therefore, seven instances of the top claim in Fig. 4i and six instance of the top claim in Fig. 4ii are generated in the safety case to address the fault models for each component in WBS respectively. For each instance of the

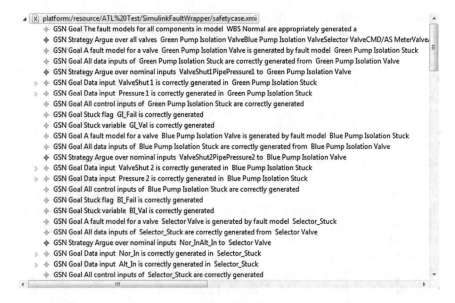

**Fig. 10** WBS safety case

fault model in WBS, two sub-claims have been generated based on the types of input. The rest of WBS safety case nodes are generated based on the WBS artifacts in the same manner. Here, the result of the generated WBS safety case is represented in a XML format and partially shown in Fig. 10.

## 5  Support of System Evolution

Artifact modification during SDLC is inevitable. As a result, safety analysis should be performed accordingly. We assume that before artifact modification occurs, a safety case has been assessed and the confidence about the whole argument structure has been accepted. Next, we study how a new modification in an artifact makes an impact on safety analysis. In order to minimize the effort for safety re-analysis, we highlight the affected nodes in a safety case. A highlighted node represents the confidence about the node being affected by the change originated from the modification in an artifact.

In this paper, we consider the modification in a nominal model since it triggers the generation of a fault model. The modification of a nominal model can have three consequences on a generated safety case shown in Fig. 11. First, a modification can result in the corresponding node in a safety case becoming invalid. For instance, if

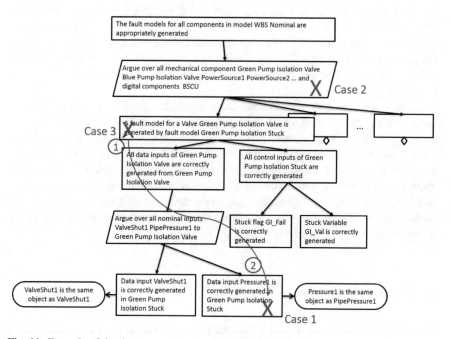

**Fig. 11** Example of the three consequences of WBS safety case

the input pump to the Green Pump Isolation Valve is changed from the Green Pump to the Blue Pump for some reason, then the sub-claim connecting it to the nominal model artifact should be affected and thus highlighted in Fig. 11(case 1) since the sub-claim requires that the input pump to the Green Pump Isolation Valve in the fault model be the same as the input pump to the Green Pump Isolation Valve in the nominal model.

The second type of modification which affects the prior argument structure of a safety case originates from adding a new element to a nominal model. In this case, the new element triggers the update of the fault model based on the failure mode for the new element. For instance, if we add a new power source component in the wheel brake system, then the Stuck_at failure mode should be added to the fault model. Thus, according to the safety case pattern in Fig. 5, a new sub-claim should be added to the strategy, arguing that all mechanical components and digital components are correctly generated by a fault model in the prior safety case. Since all the sub-claims in a safety case connecting to the strategy remain valid, we only highlight the strategy node to indicate that a new added component in a nominal model can affect the current strategy in Fig. 11(case 2).

Last, the third type of modification which affects the prior argument structure of a safety case is the deleting of an element from a nominal model when the deleted element has a failure mode and thus would affect some nodes related to it. For instance, if we assume that the Green Pump Isolation Valve is deleted for some reason from a nominal model, then the claim node arguing that the Binary_Stuck_at failure mode for the Green Pump Isolation Valve is correctly generated should be affected and thus highlighted in Fig. 11(case 3). At the same time, the deletion of the Green Pump Isolation Valve also removes the input pressure from the Green Pump, thus, the first type of modification occurs, that is, a nominal input to the valve is deleted. Then the sub-claim connecting to the nominal model is affected and highlighted.

Once a node in a safety case has been highlighted, all nodes along the path upward from the node to the root claim as well as the path downward from the node to its leaf node, i.e. the evidence node, are all highlighted. The reason we highlight these nodes is the argument structure along the path is affected by a modification of a node or nodes. Note that the root claim is always affected by a modification of any artifact used in this argument structure. However, if a path from a root claim to a leaf node does not have any highlighted node, then this path does not require any further assessment. In this way, we minimize the computation effort to perform safety analysis during system evolution.

# 6    Related Work and Conclusion

Most efforts to support software assurance for a cyber-physical system came from preventative methods. Namely, researchers proposed a variety of approaches which concentrate on the satisfaction of software assurance. These approaches can be divided into two categories. The first category is the verification-based approach which employs a verification tool to verify a quality-related attribute for a cyber-physical system. Hunt presented an approach which used ACL2r logic to represent a cyber-physical system (CPS), and then applied the ACL2r mechanical theorem prover to verify safety and progress properties in the system [12].

The second category is the testing-based approach. The disadvantage of verification-based approaches is that they face the state exploration problem in that these approaches require expansion of all possible system behaviors and may yield an exponential number of system states with respect to a system's concurrently executed activities. To overcome this disadvantage, many researchers proposed various testing-based approaches. For instance, Woehrle et al. proposed to test the conformance between an implementation and its design model [13]. The approach employed formal, state-based models for specifying expected behavior, and for representing a series of measurements of a physical quantity. They applied a timed model checker to investigate the conformance of expected and observed behavior. The measured conformance of the expected and observed behavior guided test engineers to detect errors in a cyber-physical system.

With the development of Model-Driven Engineering, model-based software development for a cyber-physical system has become a new trend in the CPS community. Pajic et al. applied the idea of a model-based development process to develop the UPP2SF model-transformation tool which aimed to automatically translate verified models in UPPAAL to models in Simulink/Stateflow which can then be simulated and tested [14]. They applied the UPP2SF tool in the design of a pacemaker to illustrate the importance of integrating modeling, verification, code-generation and testing process together for a cyber-physical system.

While all of the above effort has been made to show software assurance has been built into a cyber-physical system, it still demands some evaluative methods that can assure all quality-related attributes of a cyber-physical system have been correctly and completely derived during the SDLC. To this end, researchers employed various approaches to argue that a cyber-physical system is correctly designed and thus implemented. Panesar-Walawege et al. proposed the application of a UML profile to model a standard/guidance [15]. This UML profile is then used as a template to guide how a safety critical system can be developed and reviewed under the standard/guidance. A drawback of this approach is that the profile enforces one way to model the standard/guidance via the pre-defined stereotypes. Most systems do not however use these pre-defined stereotypes and therefore can't take advantage of the profile to enforce the development and review process in order to conform to the standard/guidance. Our work builds upon the existing work on safety case patterns and model-based approaches to safety to support the development of safety cases for cyber-physical systems.

Ayoub et al. proposed a safety case pattern for a system developed using a formal method [16]. The pattern considers whether a design model satisfies the requirements of an implementation in terms of a formal notation. Our safety case pattern generation approach can complement their pattern since both patterns target different phases of a software development process. More importantly, while Ayoub et al. did not present how to generate a specific assurance case based on their template, generation of a safety case in this paper is totally built on the model transformation technique in MDA.

Denney et al. found the heterogeneity of safety-relevant information existing in a safety-critical system and presented methods for managing the heterogeneous types of artifacts in a single argument structure [17]. The approach consists of two parts. The first part automatically assembles safety cases which are auto-generated from the application of a formal method to software, but the second part requires that safety cases from system safety analysis be manually created.

Hauge et al. proposed a pattern-based method to facilitate software design for safety critical systems [18]. Underlying the pattern-based method is a language that offers six different kinds of basic patterns as well as operators for composition. One of the important ramifications of this method is the generation of a safety case which is connected to the artifacts produced during a development process.

Denney et al. proposed a lightweight method to automatically create a safety case from a given set of artifacts [19]. Actually, their approach considered the meta-information about the system artifacts derived during a development process. For instance, they considered a table structure for hazards, and system/functional requirements. Namely, each hazards table and system functional requirements table have their own structure. Based on the table's structure, an algorithm was proposed to create a safety case for each system.

In this paper, to support the generation of an argument structure for a cyber-physical system, we consider the transition from trustworthiness to resiliency for a cyber-physical system as an example to illustrate how MDA can leverage the generation of a safety case via safety case patterns. Specially, we take the wheel brake system into account and demonstrate a safety case is generated via three safety case patterns related to the translation from a nominal model into a fault model. We further discuss how artifact modification can affect the argument structure of a safety case during system evolution.

**Acknowledgments** This project is supported by the Air Force Summer Faculty Research Program. The authors would like to thank the colleagues from AFRL and many universities across the countries for the discussion about this project.

# References

1. National Research Council: Critical Code: Software Producibility for Defense. National Academies Press, Washington, D.C. (2010)
2. CHC, Q.C.: The nimrod review (2009)
3. Kelly, T.: Arguing safety—a systematic approach to manage safety cases. Doctoral Dissertation, Department of Computer Science, University of York (1998)
4. Joshi, A., Heimdahl, M., Miller, S., Whalen, M.: Model-Based Safety Analysis Final Report. Contractor report Cecilia Haskins, Nasa Langley Research Center (2006)
5. Object Management Group (OMG). http://www.omg.org/spec/OCL/2.3.1/PDF/ (2012)
6. Eclipse's ATL. http://eclipse.org/atl/ (2015)

7. SAE ARP 4761: Guidelines and Methods for Conducting the Safety Assessment Process on Civil Airborne Systems and Equipment. SAE International (1996)
8. Leroux, D., Nally, M., Hussey, K.: Rational software architect: a tool for domain-specific modeling. IBM Syst. J. **45**(3), 555–568 (2006)
9. Hawkins, R., Habli, I., Kolovos, D., Paige, R., Kelly, T.: Weaving an assurance case from design: a model-based approach. In: IEEE 16th International Symposium on High Assurance Systems Engineering (HASE) (2015)
10. E. Fundation. http://www.eclipse.org/modeling/mdt/
11. W3School. http://www.w3schools.com/xml/dom_parser.asp (2015)
12. Hunt, W.: Modeling, verification of cyber-physical systems. In: National Workshop on High-Confidence Automotive Cyber-Physical Systems (2008)
13. Woehrle, M., Lampka, K., Thiele, L.: Conformance testing for cyber-physical systems. ACM Trans. Embed. Comput. Syst. **11**(4), 1–23 (2013)
14. Pajic, M., Jiang, A., Lee, I., Sokolsky, O., Mangharam, R.: Safety-critical medical device development using the UPP2SF model translation tool. ACM Trans. Embed. Comput. Syst. **13**(4) (2014)
15. Panesar-Walawege, R., Sabetzadeh, M., Briand, L.: Supporting the verification of compliance to safety standards via model-driven engineering: approach, too-support and empirical validation. Inf. Softw. Technol. **55**, 836–864 (2013)
16. Ayoub, A., Kim, B., Lee, I., Sokolsky, O.: A safety case pattern for model-based development approach. In: NASA Formal Methods, pp. 141–146. Springer (2012)
17. Denney, E., Pai, G.: Automating the assembly of aviation of safety cases. IEEE Trans. Reliab. **63**(4) (2014)
18. Hauge, A.A., Stølen, K.: A pattern-based method for safe control systems exemplified within nuclear power production. In Comput. Saf. Reliab. Secur. **7612** (2012) (LNCS)
19. Denney, E., Pai, G.: A lightweight methodology for safety case assembly. In: Computer Safety, Reliability, and Security, pp. 1–12. Springer (2012)

# Select and Test (ST) Algorithm
# for Medical Diagnostic Reasoning

D.A. Irosh P. Fernando and Frans A. Henskens

**Abstract** This paper presents an enhanced version of the ST Algorithm, which has been published previously (Fernando and Henskens , Polibits 48:23–29, 2013 [17]). The enhancements include improved presentation of the knowledgebase, a special bipartite graph in which the relations between a clinical feature (e.g. a symptom) and a diagnosis represent two posterior probabilities (probability of the diagnosis given the symptom, and the probability of the symptom given the diagnosis). Also, the inference step, induction, which estimates the likelihood (i.e. how likely each diagnosis is) has been improved using an orthogonal vector projection method for calculating similarities. The algorithm has been described in a more mathematical form (e.g. using sets rather than the linked lists that were used in its earlier version) mostly as a manipulation of sets by adding and removing elements that are in the bipartite graphs. The algorithm was implemented in Java, and a small knowledge base has been used in this paper as an example for illustration purpose only. The focus of this paper is on the algorithm, which is intended to give a theoretical proof that medical expert systems are achievable; the design and implementation of a knowledgebase that can be practically useful for clinical work, was not within the scope of this work.

**Keywords** ST algorithm · Medical expert systems · Medical diagnosis

D.A.I.P. Fernando (✉)
Distributed Computing Research Group, School of Electrical Engineering
and Computer Science, School of Medicine and Public Health, University of Newcastle,
Callaghan, NSW, Australia
e-mail: irosh.fernando@uon.edu.au

F.A. Henskens
Distributed Computing Research Group, Health Behaviour Research Group,
School of Electrical Engineering and Computer Science, University of Newcastle,
Callaghan, NSW, Australia
e-mail: frans.henskens@newcastle.edu.au

© Springer International Publishing Switzerland 2016
R. Lee (ed.), *Software Engineering, Artificial Intelligence, Networking
and Parallel/Distributed Computing*, Studies in Computational
Intelligence 653, DOI 10.1007/978-3-319-33810-1_6

# 1 Introduction

Attempts to create successful medical expert systems that can be used in real word settings go as far back as to the early conception of Artificial Intelligence. Even though there have been a number of efforts several decades ago that required a large scale manpower, development of clinically useful systems have been very challenging [1].

The process of medical diagnosis generally consists of two stages: (1) search for clinical findings (e.g. clinical symptoms and signs, investigations such as laboratory tests); (2) arriving at a diagnostic conclusion based on the findings of the previous stage. There is a number of models that have already been introduced for medical diagnostic reasoning. Some of these include: Parsimonious Covering Theory [2]; Information Processing Approach [3]; Process Model for diagnostic reasoning [4]; Certainty Factor model [5]; models based on Bayes Theorem [6–8] and Fuzzy logic [9–11]; scheme-inductive reasoning [12]; hypothetico-deductive reasoning [13]; backward and forward reasoning [14]; and pattern recognition [15]. The overall limitation of these approaches was that they do not adequately cover the two stages of diagnostic reasoning.

Development of effective search algorithms are essential for successful development of medical expert systems because failure to search for all relevant clinical findings that can be critical in arriving at an accurate diagnosis can adversely impact on the reliability of such a system, rendering it unusable. Based on the Select and Test model that was introduced by Ramoni et al. for medical diagnosis [16], ST algorithm was developed [17] to fulfil this need for an approach that cover the both stages of medical diagnostic reasoning adequately. However, the initial version of the ST algorithm lacked a proper method for deriving the likelihood of each possible diagnosis. When a vector of clinical features is to be compared to a diagnosis, which has been represented as a standard vector of expected clinical features, a method called orthogonal vector projection can be used to measure the similarity between these two vectors [18, 19]. Since the second stage of the clinical reasoning involves a similar problem (i.e. comparing the similarly of the vector of patient symptoms with the expected symptom vector in relation to each likely diagnosis), the ST algorithm was enhanced using the orthogonal vector projection method.

# 2 The Orthogonal Vector Projection Method

The geometric interpretation of the orthogonal vector projection method involves deriving the ratio described in Fig. 1. Deriving this ratio in a vector space of $R^n$ where $n$ is the number of clinical features consisting of a diagnosis, involves representing any given standard diagnosis criteria as a $\vec{S} = <s_1, s_2, \ldots, s_n>$ where $s_1, s_2, \ldots, s_n$ represent the expected severities of each clinical feature in a typical severe case. Then ascertaining if any given clinical case represented as a vector $\vec{X} = <x_1, x_2, \ldots, x_n>$ where $x_1, x_2, \ldots, x_n$ represent the elicited severities of

**Fig. 1** Deriving the similarity of two vectors using the vector projection method

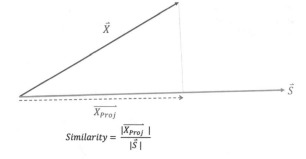

$$Similarity = \frac{|\overrightarrow{X_{Proj}}|}{|\vec{S}|}$$

each clinical feature, involves projecting $\vec{X}$ onto $\vec{S}$ resulting in the projected vector $\overrightarrow{X_{proj}}$. How the given constellation of symptoms that are embodied in $\vec{X}$ is similar to the diagnosis criteria $\vec{S}$ is determined by the ratio $|\overrightarrow{X_{proj}}|/|\vec{S}|$, which is the length of $\overrightarrow{X_{proj}}$ to the length of $\vec{S}$ as shown in Fig. 1. This ratio gives a clinically intuitive and meaningful interpretation since the similarity of diagnosis is a measure of how parallel $\vec{X}$ is to $\vec{S}$ whereas the length of $\vec{X}$ represents the overall severity of the symptoms of the case.

## 3 Modelling the Process of Medical Diagnosis

A medical diagnostic consultation generally starts with a patient presenting to a medical practitioner with a set of clinical features, which are usually called the presenting complaint. We use the term symptoms synonymously with clinical features. Thus symptoms include all clinical findings, not only what the patient reports but also findings of clinical examinations and medical investigations. In the ST algorithm, we define the set *PresentingSymptoms* which contains the individual elements of clinical features that are initially known and reported to medical practitioner by the patient. The set *PatientProfile* is defined as the set of all known and unknown clinical features presented by a patient for any given diagnostic consult, and it is associated with a vector $Q$ consisting of a quantification the severity of each symptom, $q(s)$ for all $s \in PatientProfile$ where $q(s)$ is scaled in a such a way that $q(s) \in [0,1]$. Therefore, *PresentingSymptoms* is a subset of *PatientProfile*, and the set difference, *PatientProfile-PresentingSymptoms* = $\{s|s \in PatientProfile$ AND $s \notin PresentingSymptoms\}$ represents the clinical features that are unknown at the beginning, but that need to be explored in order to have all the necessary information for arriving at an accurate diagnosis. For example, a patient may present with fever, which is an element of *PresentingSymptoms*, and $q(fever)$ represent the measured temperature scaled down to [0,1]. However, the fact that the patient has high white blood cell count is initially unknown, and can only be revealed after doing a blood test, which needs to be requested by the medical practitioner. The first stage of the diagnostic algorithm

**Fig. 2** Illustration of
diagnostic inference process
using posterior probabilities

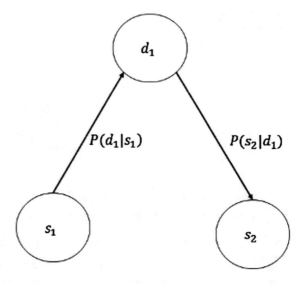

involves searching for the unknown clinical features in *PatientProfile* using cues
that are derived from the elements in *PresentingSymptoms*.

This search process can be modelled as a traversal of a special type of bipartite
graph $K_{m,n}$ which represents the entire knowledgebase with the set of vertices
$\{d_1, d_2, \ldots, d_m\}$ as all possible diagnoses and $\{s_1, s_2, \ldots, s_n\}$ as all possible
clinical features. Each pair $(d_i, s_j)$, $i = 1, \ldots, m$; $j = 1, \ldots, n$ has two edges
representing the posterior probabilities $P(s_j \mid d_i)$ and $(d_i \mid s_j)$, which are the probability of having the clinical feature $s_j$ given the diagnosis of $d_i$, and the probability
of having diagnosis of $d_i$ given clinical feature $s_j$ respectively. For example (Fig. 2),
consider that a patient is known to have a clinical feature $s_1$. The diagnostic
inference in this situation uses $P(d_1 \mid s_1)$ in determining $d_1$ as a potential diagnosis.
If $d_1$ is considered as a potential diagnosis, then it is necessary to look for other
clinical features that are associated with $d_1$ using $P(s_j \mid d_i)$. For example, the
decision if $s_2$ is to be explored in *PatientProfile* is determined by $P(s_2 \mid d_1)$.

It is obvious to the reader that when m and n are small there is no need for a
search algorithm since it is possible to go through each diagnosis and each
symptoms sequentially in a systematic way. However, if the algorithm is to cover
the entire set of diagnoses and clinical features across various medical specialities,
then m and n become large numbers, and it would not be feasible to go through
each diagnosis and each symptom sequentially in a systematic way because of the
large amount of time it would require. It also needs to note that as m and n become
large, some or most of the two posterior probabilities would be zero since each
diagnosis is not associated with each clinical feature in a clinically meaningful way.
Nevertheless, the representation of knowledgebase as a bipartite graph is preferred
because of the simplicity.

In the process of medical diagnosis, even though the $P(d_1 \mid s_1)$ is low, the
possibility of $d_1$ still needs to be ruled out, particularly if $d_1$ is associated with high

criticality (e.g. life-threatening if the diagnosis of $d_1$ is missed). Therefore, each $d_i$, $i = 1, \ldots, m$ are associated with a value of criticality denoted as $C(d_i)$.

The two main logical inferences that are involved in the search process are abduction and deduction. Abduction involves: for any given clinical features $s$ that was found in patient profile, extracting all diagnoses $d_i$ such that $P(d_i \mid s) \geq t_S$ OR $C(d_i) \geq t_C$ where $t_S$, $t_C$ are arbitrary threshold values. These diagnoses are stored in a set named *DiagnosesToBeElicited*. On the other hand, deduction involves: for each $d \in$ *DiagnosesToBeElicited*, extracting all clinical features $s_j$ such that $P(s_j \mid d) \geq t_S$, and storing them in a set named *SymptomsToBeElicited*. Afterwards, $d$ is removed from the set *DiagnosesToBeElicited* and stored in *DiagnosesAlreadyElicited* The next step, which is known as abstraction according to the ST model, involves: for each clinical features $s \in$ *SymptomsToBeElicited*, checking if $s \in$ *PatientProfile* Afterwards, $s$ is removed from *SymptomsToBeElicited* and stored in *SymptomsAlreadyElicited* Furthermore, if $s \in$ *PatientProfile* then $s$ is also stored in a set named *SymptomsFound* In essence, the ST algorithm involves recursively applying abduction, deduction, and abstraction until the search for all the relevant diagnoses and clinical features are completed.

The final step, which is known as induction involves: for each $d \in$ *DiagnosesToBeElicited* calculating the diagnostic likelihood of (i.e. how likely for patient to have) diagnosis $d$ based on the elements of the set *SymptomsFound*. The reader should not confuse the term likelihood used in this context with the concept of likelihood function used in statistics. Using the vector $S$, which contains $P(s_j \mid d)$ as elements and $Q$, which contains the quantification of clinical features $q(s_j)$, the likelihood of the diagnosis $L(d)$ is calculated as follows using the orthogonal vector projection method:

$$L(d) = \frac{S \cdot Q}{|S|^2}$$

where,

$$S \cdot Q = \left( \sum_{i=1}^{n} P(s_j \mid d) \cdot q(s_j) \right)$$

$$|S|^2 = \sum_{i=1}^{n} P(s_j \mid d)^2$$

$$q(s_j) = 0 \ if \ s_j \notin SymptomFound$$

If the diagnostic likelihood is above the diagnostic threshold value $t_C$, then $d$ is stored in a set named *Diagnoses Included*. Otherwise, $d$ is stored in a set named *Diagnoses Excluded*. The complete ST algorithm is listed in Appendix 1.

Alternative to the orthogonal vector projection method, one may consider Bayesian approach to calculate the probabilities of likely diagnoses. However, it would require various joint probability distributions, which can become not only too complicated by also tends to become less meaningful clinically.

The page limit of this paper didn't permit including the java implementation ST algorithm in this paper. However, reader may refer elsewhere for the java implementation of the previous version of the ST algorithm [17].

# 4 Implementation

In order to further describe the ST algorithm, it has been implemented in Java using a small knowledgebase. It is important to emphasise that implementation of a knowledgebase that could be of practical value would require a large scale of manpower and resources, and it was not within the scope of this research. The sample knowledgebase used here is for illustration purpose only, and may not be an accurate representation of clinical knowledge.

The sample knowledgebase consists of the following set of diagnoses:

{*Major Depression, Generalised Anxiety Disorder, Hyperthyroidism, Paechromocytoma, Anaemia, Ischaemic Heart Disease*}.

The following is the set of symptoms:

{*Depressed Mood, Loss of Motivation, Weight Loss, Fatigue, Chest Discomfort, Worrying Thoughts, Low Self-Esteem, Headache, Loss of Appetite, Hand Tremors, Hypertension, Dizziness*}.

The posterior probabilities $P(s_j \mid d_i)$ and $P(d_i \mid s_j)$ are shown below in matrix form, respectively. Please note that the symptoms represent columns while the diagnoses represent rows.

$$
\begin{bmatrix}
1.0 & 1.0 & 0.7 & 0.7 & 0.4 & 0.6 & 1.0 & 0.5 & 0.7 & 0.0 & 0.0 & 0.0 \\
0.6 & 0.5 & 0.0 & 0.7 & 0.6 & 1.0 & 0.5 & 0.7 & 0.0 & 0.6 & 0.5 & 0.7 \\
0.0 & 0.0 & 0.8 & 0.7 & 0.5 & 0.4 & 0.0 & 0.0 & 0.0 & 0.8 & 0.0 & 0.3 \\
0.0 & 0.0 & 0.0 & 0.0 & 0.0 & 0.0 & 0.0 & 0.6 & 0.0 & 0.0 & 0.9 & 0.0 \\
0.5 & 0.5 & 0.0 & 0.8 & 0.0 & 0.0 & 0.0 & 0.0 & 0.0 & 0.0 & 0.0 & 0.6 \\
0.0 & 0.0 & 0.0 & 0.3 & 0.8 & 0.0 & 0.0 & 0.0 & 0.0 & 0.0 & 0.5 & 0.6
\end{bmatrix}
$$

$$
\begin{bmatrix}
0.9 & 0.9 & 0.6 & 0.6 & 0.0 & 0.0 & 0.6 & 0.0 & 0.7 & 0.0 & 0.0 & 0.0 \\
0.0 & 0.0 & 0.0 & 0.7 & 0.6 & 0.9 & 0.4 & 0.6 & 0.0 & 0.6 & 0.0 & 0.4 \\
0.3 & 0.0 & 0.7 & 0.6 & 0.0 & 0.0 & 0.0 & 0.0 & 0.0 & 0.8 & 0.0 & 0.3 \\
0.0 & 0.0 & 0.0 & 0.0 & 0.0 & 0.0 & 0.0 & 0.5 & 0.0 & 0.0 & 0.9 & 0.0 \\
0.0 & 0.3 & 0.0 & 0.8 & 0.0 & 0.0 & 0.0 & 0.0 & 0.0 & 0.0 & 0.0 & 0.6 \\
0.3 & 0.0 & 0.0 & 0.3 & 0.8 & 0.0 & 0.0 & 0.0 & 0.0 & 0.0 & 0.4 & 0.6
\end{bmatrix}
$$

The set of diagnoses are associated with the following vector of critical values: $(0.7, 0.4, 0.4, 0.7, 0.2, 0.7)$

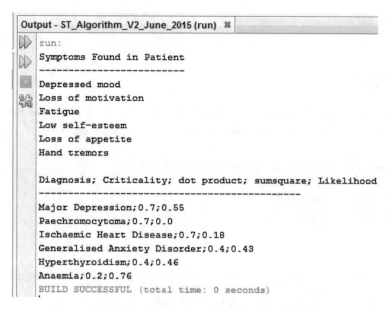

```
Output - ST_Algorithm_V2_June_2015 (run)  ⊠

run:
Symptoms Found in Patient
------------------------
Depressed mood
Loss of motivation
Fatigue
Low self-esteem
Loss of appetite
Hand tremors


Diagnosis; Criticality; dot product; sumsquare; Likelihood
----------------------------------------------------------
Major Depression;0.7;0.55
Paechromocytoma;0.7;0.0
Ischaemic Heart Disease;0.7;0.18
Generalised Anxiety Disorder;0.4;0.43
Hyperthyroidism;0.4;0.46
Anaemia;0.2;0.76
BUILD SUCCESSFUL (total time: 0 seconds)
```

**Fig. 3** Output of the ST algorithm that was implemented in java

For example, given the following patient profile:

{*Depressed Mood, Loss of Motivation, Fatigue, Low Self-Esteem, Loss of Appetite, Hand Tremors*}

with the respective quantities of each of these symptoms $(0.6, 0.4, 0.8, 0.7, 0.9, 0.6)$ and the presenting clinical features {*Depressed Mood, Fatigue*} the ST algorithm produces the following output in Fig. 3.

## 5 Conclusion

This paper has presented an improved version of the ST algorithm, which can potentially be used to develop large scale medical expert systems that cover a broader clinical knowledge across various medical specialities. However, achieving such a large scale medical expert system is a challenging task, which requires large-scale resources and manpower. Also, it requires integration of various technologies such as image processing, speech recognition, and biomedical signal processing depending on how sophisticatedly the abstraction step of the algorithm is implemented. Unless, it can be proven that the system is achievable, such a large-scale effort would a waste. The existence of ST algorithm provides the required theoretical proof that such systems are achievable. Whilst a simpler design of knowledgebase with two layers (i.e. clinical features and diagnoses) is presented in this paper, it is possible to enhance it by adding more layers(e.g. a layer of attributes that are associated with each clinical feature) as required.

# Appendix

## Inputs:

A special bipartite Graph $K_{m,n}$ which has set of vertices $d_1, \ldots, d_m$ as diagnoses and $s_1, \ldots, s_n$ as symptoms and each pair $(d_i, s_j)$ has two edges representing $P(s_j | d_i)$ and $P(d_i | s_j)$.

A presenting symptoms set, $PresentingSymptoms$
Profile of the patient, $PatietProfile$
Threshold for symptoms, $t_S$
Threshold for diagnoses, $t_C$

## Outputs:

Set of likely Diagnoses, $Diagnoses\ Included$
Set of Diagnoses excluded, $Diagnoses\ Excluded$
Set of Symptoms that were found in patient, $SymptomsFound$
Set of Symptoms that were not found in patient, $SymptomsNotFound$
**Begin**
$SymptomsFound = \emptyset$
$SymptomsToBeElicited = \emptyset$
$SymptomsAlreadyElicited = \emptyset$
$DiagnosesToBeElicited = \emptyset$
$DiagnosesAlreadyElicited = \emptyset$

For each $s \in PresentingSymptoms$ {
      add $s$ to $SymptomsAlreadyElicited$;
      add $s$ to $SymptomsFound$;
      For each $d_i$ where $P(d_i | s) \geq t_D$ OR $C(d_i) \geq t_C$ {
            add $d_i$ to $DiagnosesToBeElicited$.
      }
}
While $(DiagnosesToBeElicited \neq \emptyset$ AND $SymptomsToBeElicited \neq \emptyset)$ {
            $Abduct()$
            $Deduct()$
            $Abstract()$
      }
      $Induct()$
**End**
// End of the main program

$Abduct()$ {
      For each $s \in SymptomsFound$ {

For each $d_i$ where $P(d_i|s) \geq t_D$ OR $C(d_i) \geq t_C$ {
    If ( $d_i \notin DiagnosesToBeElicited$
          AND $d_i \notin DiagnosesAlreadyElicited$),
      Add $d_i$ to $DiagnosesToBeElicited$
    }
  }
}

$Deduct()$ {
    For each $d \in DiagnosesToBeElicited$ {
      For each $s_j$ where $(s_j|d) > t_D$ {
      If $(s_j \notin SymptomsFound$
          AND $s_j \notin SymptomsAlreadyElicited$),
        Add $s_j$ to $SymptomsToBeElicited$.
      }
      remove $d$ from $DiagnosesToBeElicited$.
      add $d$ to $DiagnosesAlreadyElicited$.
    }
  }

$Abstract()$ {
    For each $s \in SymptomsToBeElicited$ {
      If $s \in PatietProfile$, add $s$ to $SymptomsFound$.
      Remove $s$ from $SymptomsToBeElicited$
      Add $s$ to $SymptomsAlreadyElicited$
    }
  }

$Induct()$ {
    For each $d \in DiagnosesAlreadyElicited$ {

$$L(d) = \frac{S.X}{|S|^2}$$

    Where,

$$S.X = \left( \sum_{i=1}^{n} P(s_j|d).q(s_j) \right),$$

$$|S|^2 = \sum_{i=1}^{n} P(s_j|d)^2$$

If $L(d) > t_c$ add $d$ to $DiagnosesIncluded$,
    $Else$ add $d$ to $Diagnoses\ Excluded$.
    }
  }

# References

1. Wolfram, D.A.: An appraisal of INTERNIST-I. Artif. Intell. Med. **7**, 93–116 (1995)
2. Reggia, J.A., Peng, Y.: Modeling diagnostic reasoning: a summary of parsimonious covering theory. Comput. Methods Programs Biomed. **25**, 125–134 (1987)
3. Wortman, P.M.: Medical diagnosis: an information-processing approach. Comput. Biomed. Res. **5**, 315–328 (1972)
4. Stausberg, J.R., Person, M.: A process model of diagnostic reasoning in medicine. Int. J. Med. Informatics **54**, 9–23 (1999)
5. Shortliffe, E.H., Buchanan, B.G.: A model of inexact reasoning in medicine. Math. Biosci. **23** (4), 351–379 (1975)
6. Andreassen, S., Jensen, F.V., Olesen, K.G.: Medical expert systems based on causal probabilistic networks. Int. J. Bio-Med. Comput. **28**(5), 1–30 (1991)
7. Chard, T., Rubenstein, E.M.: A model-based system to determine the relative value of different variables in a diagnostic system using Bayes theorem. Int. J. Bio-Med. Comput. **24**(7), 133–142 (1989)
8. Todd, B.S., Stamper, R., Macpherson, P.: A probabilistic rule-based expert system. Int. J. Bio-Med. Comput. **33**(9), 129–148 (1993)
9. Boegl, K., Adlassnig, K.-P., Hayashi, Y., Rothenfluh, T.E., Leitich, H.: Knowledge acquisition in the fuzzy knowledge representation framework of a medical consultation system. Artif. Intell. Med. **30**(1), 1–26 (2004)
10. Godo, L.S., de Mántaras, R.L., Puyol-Gruart, J., Sierra, C.: Renoir, Pneumon-IA and Terap-IA: three medical applications based on fuzzy logic. Artif. Intell. Med. **21**(1), 153–162 (2001)
11. Vetterlein, T., Ciabattoni, A.: On the (fuzzy) logical content of CADIAG-2. Fuzzy Sets Syst. **161**, 1941–1958 2010
12. Mandin, H., Jones, A., Woloschuk, W., Harasym, P.: Helping students learn to think like experts when solving clinical problems. Acad. Med. **72**, 173–179 (1997)
13. Elstein, A.S., Shulman, L.S., Sprafka, S.A.: Medical Problem-Solving: an Analysis of Clinical Reasoning. Harvard University Press, Cambridge, MA (1978)
14. Hunt, E.: Cognitive science: definition, status, and questions. Annu. Rev. Psychol. **40**, 603–629 (1989)
15. Norman, G.R., Coblentz, C.L., Brooks, L.R., Babcook, C.J.: Expertise in visual—a review of the literature. Acad. Med. **66**(suppl), s78–s83 (1992)
16. Ramoni, M., Stefanelli, M., Magnani, L., Barosi, G.: An epistemological framework for medical knowledge-based systems. IEEE Trans. Syst. Man Cybern. **22**, 1361–1375 (1992)
17. Fernando, I., Henskens, F.: ST algorithm for diagnostic reasoning in psychiatry. Polibits **48**, 23–29 (2013)
18. Fernando, I., Henskens, F.: A modified case-based reasoning approach for triaging psychiatric patients using a similarity measure derived from orthogonal vector projection. In: Chalup, S., Blair, A., Randall, M. (eds.) Artificial Life and Computational Intelligence, vol. 8955, pp. 360–372. Springer, Switzerland (2015)
19. Fernando, I., Henskens, F.: A case-based reasoning approach to mental state examination using a similarity measure based on orthogonal vector projection. In: MICAI 2014 (2014)

# Opportunities Ahead the Future Mobile Learning

Regin Joy Conejar, Haeng-Kon Kim and Roger Y. Lee

**Abstract** Technology is having an unprecedented impact on education; its future is being shaped by current and emerging technologies that are drastically changing the way in which learning and teaching are experienced. Education is increasingly becoming individualized, customized and more accessible as a result of combining open source technology, the Internet, mobile and multi-faceted technology, virtual learning environments and learning analytic technology. This report reflects on the innovation and the complexities that are currently emerging in education as a result of these technological advancements.

**Keywords** Advance learning · Educational technology

## 1 Introduction

In the modern landscape, ensuring excellent quality of education is just as important as the typically more emphasized focus on increasing the quantity of those in education, at any level. This is where educational technology has such a substantial impact—not only does it greatly improve the interactivity, and engagement of the educational experience (quality), but it also brings with it improved accessibility and universality (quantity). Educational technology is so vitally important to these targets because it can improve them both simultaneously. These are two targets

R.J. Conejar · H.-K. Kim
School of Information Technology, Catholic University of Daegu, Daegu, Korea
e-mail: regin@cu.ac.kr

H.-K. Kim
e-mail: hangkon@cu.ac.kr

R.Y. Lee (✉)
Department of Computer Science, Central Michigan University, Michigan, USA
e-mail: lee1ry@cmich.edu

© Springer International Publishing Switzerland 2016
R. Lee (ed.), *Software Engineering, Artificial Intelligence, Networking and Parallel/Distributed Computing*, Studies in Computational Intelligence 653, DOI 10.1007/978-3-319-33810-1_7

which, policy wise, have historically been targeted independently by governments and institutions, as they have previously involved very different approaches.

Educational technology has combined these approaches, and has accelerated education's future trajectory greatly. It is an exciting time to be involved in the educational sector, regardless of your role. The revolution in Education does not just affect the way in which students learn—this particular paradigm shift reverberates significantly deeper than that. It affects the way teachers teach, the way schools are structured, the barriers between school and home life, and—perhaps on its most profound level—affects the trajectory of the entire future of humanity. The global future of mankind in these modern, changing times is uncertain, unstable, and dynamic. In order for future generations to adapt to such uncertainty, and create sustainability, it is vital that the way in which we teach them to do so can also adapt with equal dynamism. This is not the case with the "old" educational paradigms.

Technology has changed our world in ways previously unimaginable. Mobile devices permeate our daily lives, providing unparalleled access to communication and information. Looking towards the next decade and beyond, it seems clear that the future of mobile learning lies in a world where technology is more accessible, affordable and connected than it is today. However, technology alone, regardless of its ubiquity and utility, will not determine whether mobile learning benefits large numbers of people. Designing effective mobile learning interventions requires a holistic understanding of how technology intersects with social, cultural and, increasingly, commercial factors. The technology itself is undeniably important, but equally if not more important is how people use and view technology, a point that has been largely overlooked. Just because mobile devices carry a potential to, say, help improve the literacy skills of women in resource-poor communities does not mean that these devices will actually be employed towards this end. Indeed, across the world women are far less likely than men to own and use mobile devices, and in many communities women are discouraged from using mobile technology for any purpose, learning included. Mobile devices are often banned from schools and other centres of education, despite considerable and, in many instances, well-established potential to enhance learning. Such bans project a view that mobile devices are antithetical to learning, and this outlook, regardless of its factual validity, impacts the way people interact with technology. Over the next 15 years, the implementation of mobile learning projects and the pedagogical models they adopt should be guided not only by the advantages and limitations of mobile technologies but also by an awareness of how these technologies fit into the broader social and cultural fabric of communities.

Education and technology can and should co-evolve in mutually supportive ways. While people tend to think of education as perpetually lagging behind technology, there are numerous instances in which education has prompted technical innovation. For example, some historians argue that Alan Kay's 1968 Dynabook, an early prototype of the laptop computer, came into existence as a means of helping students learn through 'new media'. Kay drew on the theories of well-known learning specialists to inform the design and functionality of the device (Dalakov 2013). With the unprecedented rate of technology change, it may seem nearly impossible to imagine what mobile learning will look like a decade from now, let alone two decades. Yet exploring these questions is an important exercise, as the future will be moulded by the decisions made today. With the right social and political supports and, most immediately, mechanisms to train practitioners to design mobile learning interventions, mobile leaning has the potential to transform educational opportunities and outcomes. This report helps guide the way by highlighting issues and questions likely to steer mobile learning over the next 15 years and beyond. It begins by presenting an overview of the current state of mobile learning, describing recent mobile learning developments in formal and informal education, seamless learning and educational technology. Based on current trends, the report then makes predictions for the future of mobile learning, forecasting likely technological advances and mobile learning focus areas. Subsequent sections discuss mobile learning in light of Education for All (EFA) goals, both now and in the future, and identify the primary enablers for mobile learning as well as the main barriers to its development [1]. Finally, the report presents the broad, overarching challenges to be met over the next 15 years, in order for mobile learning to be integrated into mainstream education and impact teaching and learning on a global scale. Ultimately, the report seeks to provide policy-makers and other education stakeholders with a tool to better leverage mobile technologies in the ongoing effort to improve educational access, equity and quality for all.

## 2 Related Works

### 2.1 Game-Based Learning

Young people have played computer and online games with enthusiasm and persistence since the 1960s and 1990s, respectively. Now computer and online games are more prolific and popular than ever before. Educational institutions, as a result of having mobile learning device initiatives and cloud computing, are harnessing the same determination, enthusiasm and persistence that are brought out of students when they play games. Cardiff teacher Gareth Ritter explains how *a lot of the kids in [this] school play Call of Duty. If they fail a level they won't give up, they'll keep doing it. We've got to bring that into the classroom* (Vasagar 2012) [2]. Game-based learning seems likely to become the most effective way to teach students fundamental concepts which would have previously been learnt via repetition and written exercises.

### 2.2 Virtual and Remote Learning Platforms

New, previously unimaginable possibilities for learning environments are also resulting from the merging of our physical and virtual world. The classroom is no longer restricted to existing inside a physical educational institution; it can be anywhere the student chooses. This is the idea encapsulated by VLEs. At the forefront of this idea are virtual and remote learning platforms. As mentioned above, VLEs are educational electronic learning systems based on online models that mimic conventional in-person education. VLEs can include most learning environments from virtual learning platforms like those of MOOCs to virtual worlds like those used for game-based learning [3]. Virtual and remote learning platforms are poised to provide any student within or outside of traditional educational infrastructure with an engaging and interactive learning environment.

## 2.3   Distance Education

The wondrous developments of technology during the Industrial Revolution brought about, for the first time in history, the possibility of distance education.

These developments were particularly important in transport and communication.

It was no coincidence, writes the German scholar Peters (1973), that the first trains, the first postal systems and the first correspondence courses commenced at the same time.

The first distance educators

- separated the teacher and the learner
- separated the learner from the learning group
- used a form of communication mediated by technology

and still claimed that the essence of the education process was maintained intact.

Distance learning systems used technology to separate the learner from the teacher, and the learner from the learning group, while maintaining the integrity of the education process.

These systems attempted to replace interpersonal communication, and the inter-subjectivity between the teacher and the taught, which is the essence of the educational transaction, by an a personal form of communication mediated by technology.

In ILT (Instructor-Led-Training/face-to-face) systems in training centres or university lecture halls, this interactivity is automatically set up.

In distance learning it has to be artificially achieved by what is known as the re-integration of the teaching acts: that is, the development of excellent distance learning materials for students studying at a distance, and the creation of excellent student support services for students in their homes, or factories, or some other place not normally geared to education and training.

## 3   The Current State of Mobile Learning

Today, mobile technologies—originally marketed mainly as communication and entertainment devices—have come to play a significant role in economies and society at large. Mobile devices have impacted nearly every field, from banking to politics, and are currently being used to increase productivity in numerous sectors. As these devices become increasingly prominent worldwide, there is a great deal of excitement building around mobile learning. Students and teachers are already using mobile technologies in diverse contexts for a wide variety of teaching and learning purposes, and key educational players—from national education ministries to local school districts—are experimenting with supportive policies to promote innovative mobile learning in both formal and informal education settings. Many of the experts interviewed for this report feel that mobile learning is now on the threshold of a more systematic integration with education both in and outside of schools. Decisions made today will fundamentally influence the character of mobile learning in years to come. To help set the stage for these decisions, the following sections outline some of the most prevalent trends in mobile learning to date. These include innovations in formal and informal education, seamless learning and educational technology.

## 3.1 Formal Education

The presence of mobile devices in formal education systems is growing. Globally, two of the most popular models for mobile learning in schools are one-to-one (1:1) programmes, through which all students are supplied with their own device at no cost to the learners or their families; and Bring Your Own Device (BYOD) initiatives, which rely on the prevalence of learner-owned devices, with schools supplying or subsidizing devices for students who cannot afford them. As might be expected, the 1:1 model tends to be more common in poorer countries and regions, while the BYOD strategy is usually implemented in wealthier communities where mobile device ownership among young people is nearly ubiquitous.

### 3.1.1 Bring Your Own Device (BYOD)

One viable way achieve a 1:1 environment is to have students use the mobile devices they already own. This model, known as BYOD, is already causing a major shift in higher education and distance learning by allowing more students to access course materials via mobile technology. As mobile access and ownership increases, BYOD holds promise for learners around the world, although it may look drastically different across various regions and countries. While the strategy has been most popular in countries and communities where smartphone and tablet ownership is widespread, learners and educators have also found ways to capitalize on less sophisticated student-owned technologies. The Nokia MoMath project in South Africa, for example, uses the SMS (Short Message Service) features on standard mobile phones to provide students with access to mathematics content and support (Isaacs 2012b).

While BYOD moves the hardware costs from the school to the learner, it places additional pressure on bandwidth—a critical infrastructure consideration for mobile learning initiatives. Schools or governments implementing BYOD programmes must also have a strategy in place to provide devices to students who cannot afford them, either by buying the devices for the students or subsidizing their purchase. Further issues include security, privacy, adequate professional development for teachers, and a digital divide between students with cuttingedge devices and those with less powerful devices or none at all. For these reasons, examples of successful BYOD initiatives, particularly in primary and secondary institutions, are limited. However, as sophisticated mobile technologies become increasingly accessible and affordable, BYOD may form a central component of mobile learning projects in the future (Norris and Soloway 2011).

## 3.2 Informal Education

Mobile learning has developed, to a large extent, outside of formal education contexts, and the vast majority of mobile learning projects are designed for informal learning.

## 3.3 Seamless Learning

Seamless learning is defined as uninterrupted learning across different environments, including formal and informal settings. In the ideal seamless learning scenario, a learner opportunistically uses various kinds of technologies, capitalizing on the unique affordances of each—the mobility of a smartphones, for example or the superior keyboard on a desktop computer—to maintain continuity of the learning experience across a variety of devices and settings. Historically, there has been a significant divide between the formal learning that happens inside a classroom and the informal learning that occurs at home or in community environments. Numerous experts are investigating how mobile learning might help break down that barrier and bridge the gap between formal and informal learning.

## 3.4 Educational Technology

Recent innovations in mobile technologies have mostly centred around the creation of digital content, largely in the form of digital textbooks accessed via e-readers, and the development of mobile applications (apps) and software platforms for accessing educational resources via mobile devices.

### 3.4.1 Digital Textbooks and e-Readers

In formal education settings in the developed world, the transition to digital textbooks is one of the most established mobile learning trends. As e-readers and e-reading applications continue to improve, the experience of reading electronically is rapidly becoming more pleasurable and conducive to learning. New approaches to textbook conversion and creation are moving away from mere digital reproductions of printed text to visually rich interfaces that can include multimedia, interactive and collaborative elements (GSMA 2011).

The next generation of e-readers and tablets will offer new possibilities for teaching and learning. For example, e-books could enable a more social form of study, with a group of students collaborating to read, annotate and compare one or more texts on the same topic, each working from their individual mobile device (Sharples et al. 2012). Future e-books could exploit the tools built into mobile devices—such as voice recorder, camera, timer, GPS (Global Positioning System) locator, accelerometer, compass and tilt sensor—for exploratory learning, guiding the reader through experiments like testing the properties of light using the device's camera or sound using the recorder (Sharples et al. 2012). As tablet and e-reader technologies improve in quality and decrease in price, this movement towards digital textbooks could increase educational opportunities for learners around the world, particularly those who do not currently have access to high-quality physical materials.

### 3.4.2 Mobile Apps

Marketplaces for mobile apps have provided an entirely new distribution mechanism for content, stimulating substantial investment in software development for mobile devices. Educational apps are already experiencing significant growth in developed countries. These apps provide new tools for educational activities such as annotation, calculation, composition and content creation. A recent study found that 270 million apps linked to education were downloaded in 2011—a more than tenfold increase since 2009 (McKinsey & Company and GSMA 2012).

While a small number of educational apps are mapped to curriculum targets and designed for use in classroom or homework settings, the majority are intended mainly for informal learning (GSMA 2011). However, as more students use mobile devices in formal education settings, apps will likely become an important part of the mobile learning ecosystem. Not only are developers now able to bypass institutions and sell content directly to learners, but students, teachers and schools alike will be able to make small, incremental investments in micro-sized pieces of content. For example, rather than investing in the same textbook set or software solution for an entire classroom, school, district or country, educators will be able to choose from a variety of apps that are tailored to each individual learner, powering the personalized learning that is expected to characterize formal education in the future.

## 4   The Future of Mobile Learning

With over 5.9 billion mobile phone subscriptions worldwide, mobile devices have already transformed the way we live. But even though people around the globe rely heavily on mobile technology, educators and policy-makers have yet to tap its full potential to improve learning. The next decade and beyond could be transformational in incorporating mobile technologies in both formal and informal education to better meet the needs of learners and teachers everywhere. The following sections describe some of the technological advances most likely to impact mobile learning in the future, and highlight key focus areas in the development of mobile learning over the next 15 years (Fig. 1).

### 4.1   Technological Advances

In the next 15 years, technology will change in numerous ways that can be leveraged for education. It is important that educators understand there innovations so as to influence their development rather than simple react to it. Ideally technology and education will co-evolve, with education needs driving technological progress as well as adapting to it. Outlined below are some of the technological advances most likely to impact teaching and learning from a global perspective.

#### 4.1.1   Technology Will be More Accessible, Affordable and Functional

While unforeseen technological innovation is certain, the advances that will have the greatest impact on education will likely stem from a continuation of the current and most important trends in technology evolution—namely improved functionality, connectivity and memory at lower costs. Increased availability and penetration of 'smart' mobile devices and cloud-based services with advanced functionalities will open up a world of new possibilities for mobile learning solutions, allowing the types of initiatives that are currently happening to be replicated on a large scale. Many experts imagine a day where every learner in the world has

| FUNCTIONALITY | | MOBILITY | | |
|---|---|---|---|---|
| Computers | Laptop Computers | PDA'S Handhelds Palmtops | Smartphones | Mobile Phones |
| <———— E-LEARNING<————> | | <———— M-LEARNING<————> | | |

Fig. 1  Functionality and mobility definition of mobile learning

access to a powerful touch-screen tablet device and can afford both the hardware and the connectivity that enables fast and seamless access to the internet and/or other networks.

## 4.2 Mobile Learning Focus Areas

In the next 15 years, mobile learning will undoubtedly become more integrated with mainstream education. Just as computers are now viewed as crucial component to learning in the twenty-first century, mobile technologies will soon become commonplace in both formal and informal education, and gradually even the term 'mobile learning' will fall into disuse as it is increasingly associated with learning in a more holistic rather than specialized or peripheral sense. As the links between technical and pedagogical innovations improve, mobile technology will take on a clearly defined but increasingly essential role within the overall education ecosystem. The following sections outline the anticipated focus areas for mobile learning development in the foreseeable future.

### 4.2.1 Authentic and Personalized Learning

The current trend towards authentic and personalized learning will continue over the next 15 years with the aid of mobile devices. Mobile technology can support learners in exploring the world around them and developing their own solutions to complex problems while working in collaboration with peers under the guidance of skilled teachers. New sensor technologies on mobile phones, coupled with new visualization technologies in the classroom, will open up insights into physical phenomena that will be particularly useful for science learning. Already there are a number of applications that use the image-capturing capabilities of smartphones and other mobile devices to, for example, help students studying botany identify different types of trees and plants they encounter in their day-to-day lives (Leafsnap 2011). In recent years, much research has focused on the use of mobile data collection tools in epidemiology—such as Imperial College London's (2013) EpiCollect application and Nokia's (2012) Nokia Data Gathering project—which allow for the collection and real-time analysis of new kinds of data sets.

The personalization features of mobile technologies will allow learners of differing abilities and at different stages of development to progress at their own pace. Learning technologies that use artificial intelligence (AI) will become more widespread in education and will be increasingly available on mobile devices. As this is an emerging field, initial uses of AI in mobile learning in the coming years may focus on relatively simple or straightforward activities. Educators will need to ensure that this is balanced by personalized interventions that support more complex and multidimensional opportunities for learning. This will enable the development of new forms of personalized support for mobile learners. The effective design, development and implementation of personalized learning strategies require vast resources and massive investments from national education ministries. Yet this investment is likely to be worthwhile, as personalized learning holds the potential to fundamentally transform models of teaching and learning and make education more relevant, engaging, authentic and effective for students everywhere.

### 4.2.2 Mobile Programming

Over the next 15 years, students will not just use mobile devices to assist their education but will learn to program the devices themselves, developing, building and tailoring mobile applications to suit their individual desires and needs. In the process they will learn about computational thinking—the key concepts underlying approaches to programming and problem-solving—and gain vital skills for participating in the twenty-first century global economy. Indications of this trend can be seen through the emergence of mobile development labs (or tech hubs) across Sub-Saharan Africa (BongoHive, n.d.), and a recent focus on increasing the number of female software developers through coding communities such as AkiraChix (2011) in Kenya. In Europe, the rise of mobile programming in education is evidenced by the growing popularity of mobile applications for social change, such as Apps for Good (2012); projects that support young people's coding skills, including CoderDojo (2012); and cheap computing alternatives, like Raspberry Pi (n.d.). The challenge for policymakers is to maintain the current level of excitement around new programming opportunities, and to encourage the integration of mobile programming with formal education, not only in the field of computer science but also in the wide range of disciplines in which computational thinking plays a role.

## 5    The Future of Higher Ed and Its Impact on e-Learning

## 5.1    Trends Currently Affecting Teaching, Learning, and Creative Inquiry in Higher Education

1. People expect to be able to work, learn, and study whenever and wherever they want to.
2. The technologies we use are increasingly cloud-based, and our notions of IT support are decentralized.
3. The world of work is increasingly collaborative, driving changes in the way student projects are structured.
4. The abundance of resources and relationships made easily accessible via the Internet is increasingly challenging us to revisit our roles as educators.

5. Education paradigms are shifting to include online learning, hybrid learning and collaborative models.
6. There is a new emphasis in the classroom on more challenge-based and active learning.

## 5.2  Important Constraints and Challenges

1. Economic pressures and new models of education are bringing unprecedented competition to the traditional models of higher education.
2. Appropriate metrics of evaluation lag the emergence of new scholarly forms of authoring, publishing, and researching.
3. Digital media literacy continues its rise in importance as a key skill in every discipline and profession.
4. Institutional barriers present formidable challenges to moving forward in a constructive way with emerging technologies.
5. New modes of scholarship are presenting significant challenges for libraries and university collections, how scholarship is documented, and the business models to support these activities (Fig. 2).

It is not surprising, therefore, that trainers, lecturers, distance education providers and teaching institutions at all levels are increasingly using the Web as a medium for delivery. Specifically and practically this study will map the evolution from the wired virtual learning environment of today, to the wireless learning environment of tomorrow.

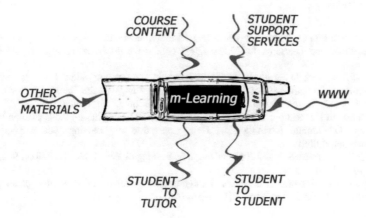

**Fig. 2**  Mobile learning environment of tomorrow [4]

# 6 Conclusion

This report has focused on how technology is impacting the future of education. The impact of technology on education, however, is not reserved for the future; technology is currently innovating the ideas and methods of education. It seems to be a very exciting time to be a student, but the best thing about the technology discussed in this report is that it allows all of us to be a student, anywhere, at any age and at any time. There's no doubt about the fact that the realm of higher education worldwide is going to undergo a vast transformation. With newer and better technology becoming increasingly affordable, classrooms the world over are evolving. Through better understanding and utilization of these incredibly powerful new revelations in educational technology, we can prepare future generations for whatever may lie ahead. This shift is placing students as independent learners, rather than teachers as instructors, at the forefront of delivering education. Teachers are increasingly becoming information guides rather than educational instructors. Consequently students are being encouraged, by these technological advancements in education, to take a more active role in their own education.

**Acknowledgments** This research was Supported by the MSIP (Ministry of Science, ICT and Future Planning), Korea, under the C-ITRC (Convergence Information Technology Research Center) support program (IITP-2015-H8601-15-1007) supervised by the IITP (Institute for Information & communication Technology Promotion).

This research was also supported by the International Research & Development Pro-gram of the National Research Foundation of Korea (NRF) funded by the Ministry of Scinece, ICT & Future Planning (Grant number: K 2013079410).

# References

1. Ahonen, M.: Project manager, hypermedia laboratory, university of tampere, Finland mobility, Accessibility and Learning. In: Mlearn 2003 conference on Learning with mobile devices (2003)
2. Attewell, J.: Mobile communications technologies for young adult learning and skills development (m-learning) IST-2000-25270. In: Proceedings of the European Workshop on Mobile and Contextual Learning. The University of Birmingham, England (2002)
3. The Future of E-Ducation: The Impact of Technology and Analytics on the Education Industry
4. Keegan, D.: Mobile Learning: The Next Generation of Learning. Distance Education International (2005)
5. Churchill, D.: Towards a useful classification of learning objects. Educ. Tech. Res. Dev. **55**(5), 479–497 (2007)
6. Design-Based Research Collective: Design-based research: an emerging paradigm for educational inquiry. Educ. Res. **32**(1), 5–8 (2003)

# A Design of Context-Aware Framework for Conditional Preferences of Group of Users

Reza Khoshkangini, Maria Silvia Pini and Francesca Rossi

**Abstract** Due to the dependency of users' preferences, which change over time, there is a need to generate a recommender framework that can handle users' conditional preferences. Since the most existing context-aware frameworks have been created for individual users, until now, little academic attention has been paid to the conditional users' preferences in group context-aware recommender systems. In dynamic domains, users' preferences are continuously affected by the domain entities, that is any object acting as service providers (e.g., restaurant, weather, users, etc. could be the entities in a restaurant selection guide). Hence, in this study we propose a context-aware framework that provides service(s) according to the current context of entities and the current users' preferences, which are naturally conditional. In addition, our framework is self-adaptive because it can adjust its own behavior (that is, the service it provides) based on the current context of entities. The Hyperspace Analogue to Context (HAC) model is used to abstract and represent the multi-dimensional entities' context to the system. The main goal of the proposed framework is to keep up at high level of satisfaction degree of group of users in a dynamic domain. We conclude this paper by simulating and evaluating our framework on a concrete scenario.

**Keywords** Context-aware recommender system · User preferences · CP-nets

R. Khoshkangini (✉)
Department of Mathematics, University of Padova, via Trieste 63,
35121 Padua, Italy
e-mail: khosh@math.unipd.it

M.S. Pini
Department of Information Engineering, University of Padova,
via Gradenigo, 35131 Padua, Italy
e-mail: pini@dei.unipd.it

F. Rossi
IBM T. J. Watson Research Center, USA (on Leave from University of Padova),
Yorktown Heights, NY, USA
e-mail: frossi@math.unipd.it

© Springer International Publishing Switzerland 2016
R. Lee (ed.), *Software Engineering, Artificial Intelligence, Networking
and Parallel/Distributed Computing*, Studies in Computational
Intelligence 653, DOI 10.1007/978-3-319-33810-1_8

# 1 Introduction

Recommender systems (RSs) have been widely used to personalize and represent information in various applications for over a decade. Such systems are divided into three main methods: *Collaborative Filtering (CF), Content-Based (CB)* and *Hybrid* (CF and CB) [14]. Content-based systems try to recommend items or services which users have rated in the past. Collaborative filtering systems provide recommendation according to the other users selections, and Hybrid systems are the combination of these two methods. Generally, these systems collect users information and provide services to fulfill the users' request [7, 15]. For example, a traditional recommender system works based on the particular knowledge of a user's preferences over the set of items, so the input of such systems is based on *user, item* and *rating*.

The next generation of RSs which are known as *context-aware recommender systems* (CARSs), are built based on *user, item* and *the context of user*. Context is basically defined as *any information that can be used to characterize the situation of an entity* [8]. From the workflow structure's point of view, CARSs have been categorized into two main methods: *Recommendation* via *context driven querying and search* and *recommendation* via *contextual preferences elicitation and estimation*. Recommendation via context driven querying and search is widely used in mobile and tourist recommender systems. Such systems use context (e.g., interest, local time, location, etc.), to query or search the user's preferences in a specific repository of resources (e.g., movie or restaurant). Thereafter, the systems try to provide the best service to users.

In contrast with the first method, in which the system employs the current users' context as a query or a preference to search for some services, the second method tries to model and learn users' preferences by observing the users' interactions with the system. Although such systems reduce the complexity of service selection tasks, they have difficulties to completely capture and understand the users' context due to the high-dimensionality of users' context and the lack of representation techniques to the system. In addition, dependency of users' preferences on the other entities' context make the users change their preferences over time. Hence, there is a need to generate an aware framework that can distinguish and consider the changes in context of entities to provide service(s).

In contrast to the most existing context-aware frameworks which are constructed for individual users, in this paper we introduce a context-aware framework which is able to provide service(s) (e.g., providing the best restaurant) according to the group of users' preferences and the other entities' context. Due to the complexity of the contextual information (consecutiveness and multi-dimensional), we used Hyperspace Analogue to Context (HAC) to abstract, handle and represent the multi-dimensional context into the system.

Users naturally use qualitative preferences rather than quantitative preferences. Moreover, users select items or features in their daily life depending on some conditions (e.g. if it's raining I prefer Italian food to Chinese food) [22]. Hence, in our context-aware framework, we used CP net in HAC model which is the most suitable way for representing qualitative and conditional preferences. To the best of our knowledge, there is no a specific study in context-aware framework that deal with different context and group of *users' conditional preferences* to provide service(s). In this study we use Majority Rule [23] mechanism to aggregate group of users' conditional preferences.

The rest of the paper is organized as follows. We discuss related studies on context and context-aware frameworks in Sect. 2, then in Sect. 3 background is reported. We define the proposed framework in Sect. 4. In Sect. 5 the recommendation module is provided. Simulation part is explained in Sect. 6 and followed by the conclusion and future work in Sect. 7.

## 2  Related Work

Context-Aware recommender systems try to recommend a set of services for a user or a group of users by considering context. Definition of context differs from one application to another, which is basically defined as *any information that can be used to characterize the situation of an entity* [8]. Simen et al. in [25] introduced a prototype of group CARS for concerts. They used Time and Location as the user's context to recommend a proper concert by implementing 3 different CF algorithms such as K-nearest neighbor, matrix factorization and Hybrid method, consisting of the combination of both methods. Palmisano and his colleagues [18] introduced a hierarchy of contextual information with multi dimensions $K = (k^1, k^2, \ldots, k^q)$ in their system, such that each dimension could have sub-dimensions such as time, location, etc. In [1] every attribute or feature is defined as a dimension (e.g., time, location) using OnLine Analytical Processing (OLAP). E.g., they considered *User* × *Item* × *Time* and defined a rating function $R$ such that they can specify how much user $u$ prefers item $i$ at time $t$. Adomavicius and his colleague in [2] defined three paradigms such as *Contextual pre-filtering*, *Contextual post filtering* and *Contextual modeling* for context-aware recommender systems which work based on *recommendation* via *contextual preferences elicitation and estimation model*. Baltrunas and Amatrian [5] introduced a method (micro-profile) in which the user's profile is divided into several sub profiles. Each sub profile represents a specific context of user. Hence the prediction's procedure acts based on these sub profiles (not a single user model). Oku and his colleagues [17] used the modeling method to incorporate additional contextual dimensions (e.g., time, weather and companion) in recommendation space, in which they used machine learning techniques (SVM) for preference classification to provide services.

## 3  Background

### 3.1  Hac

Hyperspace Analogue to Context (HAC) is a formal method to define multi-dimensional context in a Space.[1] The method is inspired by Hyperspace Analogue to Language (HAL) [16] and is used in a context-aware system to define multi-dimensional context in a smart home [20]. Rasch and his colleague introduced HAC to model services and users' preferences in their pervasive system (smart home) such that they defined each entity with multi context dimensions $\mathbb{H} = \langle \mathbb{D}_1, \mathbb{D}_2, \ldots, \mathbb{D}_n \rangle$. Each dimension $\mathbb{D}_i$ denotes a type of data, in which every dimension is a meta-data to describe the entity. They defined several *context points* for an object $o$, $c^o = \langle d_1, d_2, \ldots, d_n \rangle$, where $d_i \in \mathbb{D}_i$. *Context scope* ("C") in this model is denoted a subspace in $\mathbb{H}$, $C = \langle D_1, D_2, \ldots, D_n \rangle$, where $D_i \subseteq \mathbb{D}_i$, such that a context scope confines the value sets for the dimensions. For example, to specify condition on temperature [20–25 °C] in the smart home. In HAC the geometrical structure has been used to define and characterize the context relations, which dramatically increases the performance of the system in reasoning situations. The model is considered as an alternative approach over ontology-based models which have the symbol grounding problem [10]. HAC is also able to effectively capture the continues context from the different sources [20].

### 3.2  User Preferences, CP-Net and Sequential Majority

Preferences are common user's behaviors that change based on the context of dependent entities in a Space. Due to the conditional user's preferences such that a user prefers some items if some other items' values are determined, we use the conditional preferences network (CP-net) to model users' conditional preferences. CP-net [6] is a graphical model to represent the conditional and qualitative preference relations between variables. Assume a set of variables $V = \{X_1, \ldots, X_n\}$ with finite domains $D(X_1), \ldots, D(X_n)$. For each variable $X_i$, each user notates a set of parents $P(X_i)$ that can effect his preferences over the value of $X_i$. So this defines a dependency graph such that every variable $X_i$ has $P(X_i)$ as its immediate predecessors. They are sets of ceteris paribus preference statements (CP-statements). For instance, the CP-statement "I prefer romance movie to action if it is rainy." asserts that, given two genres that differ only in the condition of weather.

Sequential rule says that in every step, system elicits the users' preferences over the variable $X_i$ (it starts with independent variables, and then dependent variables).

---

[1]In this study, *Space* refers to a domain where all entities have dependencies. For example, in the space of selecting a restaurant, users, road, restaurants and weather have relations that can influence on users' preferences.

And then Majority rule is used to compute the chosen value for the variable $X_i$ [12]. For the features (variables), which are dependent on the other features, we select the preference ordering that relates to the assignment of the previous features [13]. The majority rule points that one outcome (a value of a variable) $x_j$ is majority better than another $x_j'$ ($x_j \succ_{maj} x_j'$) such that $|S_{x_j \succ x_j'}| > |S_{x_j \prec x_j'}| + |S_{x_j \rhd \lhd x_j'}|$ [23].

## 3.3 Eliciting User Preferences

Users preferences can be elicited *Explicitly* and *Implicitly* [4, 9]). For example, in traditional RSs users are needed to explicitly express their preferences into web-pages or forms using text-boxes, check-boxes, etc. (which is known as active learning) as a request for giving services. In contrast, in context-aware recommender systems users do not need to express their preferences directly. Such systems use different techniques (e.g., machine learning techniques [24]) to analyze, model and extract users' preferences from the interactions between users and systems in an Implicit manner. Employing these techniques need a significant expert knowledge to find the features and their dependencies which sometime is impractical. It needs to be mentioned, how extracting users' preferences from their context are beyond the scope of this paper.

# 4 System Framework

## 4.1 Case Study

To facilitate the presentation of the paper, let us consider a celebration scenario as an illustrative example throughout this paper. *Bob and his wife (Alice) have decided to celebrate their anniversary. They also have decided to invite some of their common friends to the celebration. They have difficulties in selecting a proper restaurant since they should consider their preferences w.r.t the context (location, price, traffic, cuisine, quality), as well as their friends' preferences w.r.t the context to select the best restaurant. Bob and Alice have some preferences in selecting food and restaurant. E.g., In one hand, Alice prefers Italian restaurant to Chinese with high quality. On the other hand Bob prefers (unconditionally) a restaurant in a location close to where they are located. In addition, they need to take into account their friends' favorite cuisines in order to select a proper restaurant. Furthermore, they are in a dynamic domain that any change in context of an entity (e.g., restaurant in the Space) may change the users' preferences. Therefore, they need to make sure that the restaurant can satisfy all the participants' requests under considering all aspect of the difficulties.*

Since the current CARS consider only the context of users to provide a service, there is a need to generate a framework which takes into account the current context and *conditional preferences* of group of users to fulfill the request of all the involved users. This is a challenging task that how much a recommender system can fulfill user(s) preferences. In addition, this task brings up certain challenging questions, such as how to better utilize context, how to interact with context aware recommender systems to make recommendation more flexible, and how to create an intelligent CARS. In order to answer these questions, we need a context-aware framework that can identify any changes in context (on-the-fly), recognize dependencies between all entities in the Space, and deal with multiple context and preferences of users to recommend the best service to the group of users. In other words, we need a self adaptive framework that can adopt its behavior (that is, a service it provides) based on the current context of the entities in the specific Space.

To reach the aforementioned goals, we proposed a framework that takes into account the current context of entities and users' conditional preferences to provide services to group of users. If the current context of entities change, which are the most relevant context, the system will recommend another service to users. For instance, if at time $t_1$ system provides restaurant $A$, and at time $t_2$ one of the most relevant context of restaurant changes (e.g., availability of restaurant from *Open* changes to *Close*), the system recognizes this change and introduce another restaurant based on the preferences of users. The most relevant context are the context which have more influence on users' preferences in a specific domain (e.g., in *Restaurant Recommendation* Space, the most relevant context dimensions and entities could be *road* as an entity: traffic as context dimension of road; *restaurant* as an entity: quality, price and location of restaurant as types of context; *user* as an entity of the Space: location of user, mood, etc. as the types of the user's context.). The most relevant context will be selected based on the users' interactions with systems using different techniques e.g., Pre-filtering [1].

As users would prefer to express conditional preferences to select item(s) or service(s), users' preferences are elicited (learning users' conditional preferences using conditional decision tree [11]) as CP-nets to request services. Then the current context of each user are examined with the collected models of users' preferences to obtain the current users' conditional preferences. Thereafter, the system generates a set of profiles that denote the current preferences of users. Since the framework provides service(s) to group of users, there is a need to aggregate all the users' preferences to get an overall preferences as the group's preferences. In this framework, based on the conditional preferences of users, we use Sequential Majority rule to aggregate users' conditional preferences to obtain the group's preferences. The system hereby produces a profile that contains the current group's preferences to send to the service discovery. Consequently, the recommendation module provides service(s) based on the current preference of the group and the available resources. All the procedures of the recommended system is shown with Algorithm 1. In the following steps entities and their context dimensions are defined in HAC model.

**Input** : Context of all Entities          ▷ e.g., users, available services, etc.
**Output**: Service to provide to Users

---

**begin** Main Algorithm
    $H$ = Space ;
    $HAC$ = Model of context representation into the system;
    $ContextChangeVar$ = Context variable;
    **for** *each entity $h_i \in H$* **do**
        Context $\leftarrow$ obtaining context (by different sensing) ;
        $HAC$ $\leftarrow$ Abstract (Context/ContextChange) ;
        **for** *each entity $h_i \in H$ in HAC* **do**
            ContextChangeVar $\leftarrow$ *Call* Context Change Algorithm ($C^h$) ;
        **end**
    **end**
    **for** *each entity $h_i \in H$* **do**
        **if** $h_i == User \in H$ **then**
            run data-mining process ;
            CP-nets models $\leftarrow$ learn users' preferences ;
            $P^{h_i} \leftarrow$ Eliciting current users preferences ($C^{h_i}$),          ▷ see Definition 6.
        **else**
            return 0 ;
        **end**
    **end**
    $P_c^G \leftarrow$ *Call* Aggregation Algorithm $\forall$ ($p^{h_1}, ..., p^{h_n}$)          ▷ see section 4.3 ;
    Service $\leftarrow$ Call Service Discovery Algorithm ($P_c^G$, Context, Available Service) ;
**end**

---

**Algorithm 1:** Main Algorithm

## 4.2 Context Integration and Abstraction to HAC

We use and extend HAC to model the entities in our context-aware framework. In addition, we model the users' conditional preferences in HAC. The following processes must be done in order to convert the low level context into the high level (meaningful) context: Context determination (specifying the most relevant context which have high influence on users' preferences in the domain), finding all possible events for a specific dimension, identifying the types of values (e.g., numerical, nominal, etc.), extracting all different values and integrating the different context in a Space using machine learning approaches [24]. We define the proposed framework with the following definitions:

**Definition 1** (*N-entity in a Space*) Space is a set of different entities $H = \langle h_1, h_2, ..., h_n \rangle$, $h_i \in H$, where all entities have correlations. E.g., in a Restaurant Recommendation Space $H$; users, road, restaurant, etc. could be the entities.

**Definition 2** (*N-dimensional HAC*) An n-dimensional HAC is an entity $h_i = \langle D_{i1}, D_{i2}, \ldots, D_{in} \rangle$, where each dimension $D_i$ is a type of context. In HAC, dimensions are different attributes or context that describe an entity. These can be e.g., location or status of an entity (e.g., is restaurant open or close). Thus, each entity is characterized by several different context dimensions. In this study only the most relevant context dimensions are used (with binary values for simplicity, however we generalized the binary values into non-binary values in the simulation section) in the service discovery.

**Definition 3** (*Context Point*) Context point of an entity $h_i$ in Space $H$ is $C^{h_i} = \langle v_1, v_2, \ldots, v_n \rangle$, where, $v_i \in D_j; i = \{1, \ldots, n\}$. The context of an entity is defined as a point in HAC. For instance, in our scenario, the different context point of a restaurant could be $\langle v_{location} = streetX, v_{time} = 9:00, v_{availability} = open \rangle$ or could be $\langle v_{location} = streetX, v_{time} = 10:00, v_{availability} = close \rangle$ in Space $H$.

**Definition 4** (*Context Range*) In context range that is a sub context of full $H$, we specify a binary value $\langle v_i, v_i' \rangle$ for each context dimension. E.g., in $h_{Restaurant} = \langle d_{cuisine}, d_{queue}, d_{time}, d_{available}, d_{quality} \rangle$, the range of $d_{cuisine} = [Chinese, Italian]$, $d_{queue} = [short, long]$, $d_{time} = [9:00, 10:00]$, $d_{available} = [open, close]$ and $d_{quality} = [low, high]$. Therefore, we define $v_i \in d_i \Leftrightarrow \forall i, d_i \in h_i$. It needs to be mentioned that we use non-binary ranges (3 values) in the simulation.

**Definition 5** (*Adaptation Context Change*) Context change $\times \Delta C = \langle \Delta d_1, \Delta d_2, \ldots, \Delta d_n \rangle$, is a function to find the changes between e.g., restaurants' current context and past context points. $\Delta d_i$ points the new value for a context dimension, $d_i' = d_i \times \Delta d_i$. $d_i$ does not change, if $\Delta d_i = \emptyset$. Context change $\Delta c$ plays a crucial role in our framework, since the function keeps the framework updated to provide service(s) based on the current context of entities. Any change in context could influence on the users' preferences. For instance, in the context of Restaurant, $h_{Restaurant} = \langle d_{cuisine}, d_{queue}, d_{time}, d_{available} \rangle$ we have $\langle d_{cuisine} = Italian, d_{time} = 9:00, d_{queue} = short, d_{available} = open \rangle$. If the time changes from $t_1$ to $t_2$, the range of queue's status and the availability of restaurant may change. If context point changes from $v$ to $v'$, the context of restaurant will be changed to $\langle d_{cuisine} = Italian, d_{time} = 10:00, d_{queue} = long, d_{available} = close \rangle$. c = $\langle Italian, 9:00, short, open \rangle \rightarrow \Delta$ c = $\langle Italian, 10:00, long, close \rangle$.

Note: In this function, we use certain thresholds or sampling intervals to avoid in vain service discovery running [21]. Although, any changes in the entities' context can effect on the user preferences, some changes are trivial to be considered. E.g., small change in temperature (1 °C) can not influence on the user's preferences, it can trigger and run the service discovery procedure which is costly (e.g., process, etc.). It is worth pointing out, this function runs for every individual entities which may lead to change the overall preferences of the group. Hence, context determination helps to determine the important context dimensions that needs to be considered in particular Space. As you can see in Algorithm 2, if a change happens in any dimensions of each entity, new context dimensions ($C'^{h_i}$) will be returned to the system.

---

**Input** : $C^{h_i}$,          ▷ context of $h_i$ entity
      threshold ,        ▷ constraint for certain context
**Output**: $C_c^{h_i}$         ▷ current context of $h_i$ entity

---

**begin** Context Change Algorithm
    $d_i$ = a type of context dimension ;
    $C'^{h_i} = C^{h_i} \times \Delta C^{h_i}$;
    **while** *not at end of context $C'^{h_i}$* **do**
        read all the dimensions ;
        **if** $d_i' = d_i[threshold]$ **then**
            return 0 ;
        **else**
            $C^{h_i} = C'^{h_i}$;
        **end**
    **end**
**end**

---

**Algorithm 2:** Context Change Algorithm

---

**Definition 6** (*CP-net in HAC*) According to the definition of CP-net, we use CP-net in HAC such that, the set of context dimensions (in this study we represented context dimensions (*CDs*) as nodes) $\mathbb{D} = \langle D_1^{h_1}, \ldots, D_n^{h_1}, D_1^{h_2}, \ldots, D_k^{h_2}, \ldots, D_1^{h_m}, \ldots, D_l^{h_m} \rangle$ are defined as the features of entities in the Space $H$. Each entity ($h_i$) has the different set of *CDs* $\langle D_1, D_2, \ldots, D_n \rangle$ in which every *CD* has the binary ranges $d, d'$. For each context dimension $D_i$, there is a set of parent feature $P(D_i)$ that can influence on preferences over the values of $D_i$. So this defines a dependency graph such that every node $D_i$ has $P(D_i)$ as its immediate predecessors. Each context dimension *CD* is annotated with a conditional preference table (CPT) which specifies the user's preferences over a context given the values of all parent context. Let consider context dimensions $\mathbb{Z} = \{D_1^{h_i}, D_2^{h_i}, D_3^{h_j}, D_4^{h_k}\}$ to be a partition of $\mathbb{D}$ with binary ranges containing $d$ and $d'$ (which are the names of the context points, and for each dimension $\forall i, \ d_i, d_i' \in D_i$). Therefore, the given CP-net is a graph $G = \{(D_1^{h_i}, D_3^{h_j}), (D_2^{h_i}, D_3^{h_j}), (D_3^{h_j}, D_4^{h_k})\}$ with the following CP-statements: $d_1 \succ d_1'$, $d_2 \succ d_2'$, $(d_1 \wedge d_2)$: $d_3 \succ d_3'$, $(d_1' \wedge d_2)$: $d_3' \succ d_3$, $d_3$: $d_4 \succ d_4'$. Here, statement $d_1 \succ d_1'$ represents the unconditional preference for dimension $D_1^{\in h_i} = d_1$ over dimension $D_1^{\in h_i} = d_1'$, while statement $d_3$: $d_4 \succ d_4'$ states that $D_4^{\in h_k} = d_4$ is preferred to $D_4^{\in h_i} = d_4'$, given that $D_3^{\in h_j} = d_3$. To this end, the following orders are obtained from the CP-statements: $d_1 d_2 d_3 d_4' \succ d_1 d_2' d_3 d_4' \succ d_1' d_2 d_3' d_4 \succ d_1' d_2' d_3' d_4'$. According to the above description of CP-net in HAC, we define user preference in $H$,

$P^h = \langle c : p \rangle$, such that $c = \langle \vee \neg \wedge \rangle$ and $p = \langle \succ \rangle$. Hence, the refined profile of user's preferences (we consider only the most preferable one) in HAC is $P^{h_{user}} = \langle d_1^{h_i}, d_2^{h_i}, d_3^{h_j}, d_4^{h_k} \rangle$ over $\mathbb{Z}$.

## 4.3　Preference Aggregation Module

Since CP-net is used to model the users' conditional preferences, there is a need to use a suitable mechanism that can handle the possible cases between users' preferences in a group. For instance, conflicting within users' preferences in a group (e.g., Bob prefers Chinese and Alice prefers Italian restaurant), dependency between context dimensions (e.g., Bob's preference for Chinese restaurant is dependent on restaurant's quality), incomplete preferences (e.g., Alice's preferences in the scenario), etc. There are several methods to aggregate such CP-nets in combinatorial domains which every method has its own strengths and weaknesses [23]. In this framework we use *Sequential Majority* rule to aggregate users' conditional preferences, such that in each step we consider one context dimension $D_i$ and then we calculate the number of users who prefers $d_i$ over $d_i'$ and vice versa. The most preferable context dimension will be selected based on $|S_{d_j \succ d_j'}| > |S_{d_j \prec d_j'}|$ (see Algorithm 3). Due to the page limit, we omit the detailed process of Sequential Majority mechanism. Since we use a concrete scenario there may be missing information within some context dimensions (e.g., price for both Bob and Alice, quality for Bob, location for Alice, etc.). Instead of considering indifferent on those context that we do not have information, in this aggregation mechanism we use CF technique to find the preference of other users who have the same context dimensions on those context as unconditional preference. Even though using CF increases the probability in conflicting users' preferences, it helps to increase the satisfaction of users by providing the same preferences which the system could not elicit from the users interactions' information with the systems.

---

**Input**　: $N$ users profiles $P^{h_1}, ..., P^{h_n}$
**Output**: $P_c^G$, a profile as a group profile

---

**begin** Preferences Aggregation Algorithm
　　　Run Sequential Majority rule on profiles $(P_1, ..., P_n)$ ;
　　　List preferences from the most to the less preferred ;
　　　*K-top preferences* ← Select top preferences as group;
　　　$P_c^G$ ← K-top preferences ;
**end**

---

**Algorithm 3:** Preferences Aggregation Algorithm

# 5 Recommendation Module

## 5.1 Service Discovery and Context Matching

In this module the algorithm tries to match the best service with the obtained group preferences. We use k-Nearest Neighbor (kNN) algorithm to find the best restaurant with respect to the users' preferences. In context matching, we are not only looking for full matches. For instance, the users' request (as a group preferences) has $n$ dimensions, and the service can only fulfill $k(where\, k < n)$ dimensions of the users' request. Thus the system worsening flip to the context (items or other values) which are least preferred (in group preferences) in the group. The system therefore can deliver services based on the second or the third most preferable group's preferences. We use *Cosine Similarity* [19] method to find the best service that has the minimum distance with the group's preferences. Algorithm 4 shows the structure of the service discovery module.

**Input** : $P_c^G, C_c^G$, Available Service ($AS$)
**Output**: Top Service ($S$), Satisfaction Score ($SS$)

---

**begin**
    *Similarity* = a matrix of all similarities values between users preferences and available services ;
    *Vectors* = set of different vectors ;
    *Vectors* ← convert ($P_c^G, C_c^G$ and AS)
    **for** *each Vector* $\in P_c^G$ **do**
        **for** *each Available Services* $\in AS$ **do**
            *Similarity* ← Cosine Similarity ($V_i^{p \in PG}, V_j^{as \in AS}$) ;
        **end**
    **end**
    *Similarity* ← sort(*Similarity*) ;
    *Top-Service* ← Select K top service (*Similarity*) ;
**end**

---

**Algorithm 4:** Service Discovery Algorithm

## 5.2 Proposed Framework's Workflow

The structure of the proposed framework is shown in Fig. 1. such that the context of different entities will be elicited by different types of sensing at $t_1$. Then, entities' context are abstracted to HAC model using the afford mentioned methods. Users current preferences are elicited (as it mentioned above, based on the users' interaction with system and the current context of different entities in the Space), and

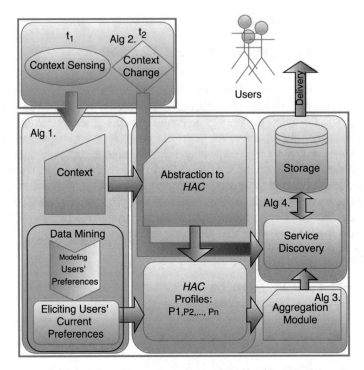

**Fig. 1** Architecture of the proposed context-aware framework, which consists of the following modules: *Sensing* part to obtain different types of context; *Data-mining* part that is responsible for learning and modeling users' conditional preferences; In *HAC* all the entities' context are abstracted to HAC model in different profiles; and *Aggregation* module is in charge of aggregating users' conditional preferences to send to the *Service discovery* module to find the best service w.r.t. the users' requests (or preferences)

then sent to the aggregation module as users' preferences profiles $(P_1, P_2, \ldots, P_n)$. In this step, aggregation module aggregates all the users' CP-nets into one or set up preferences (k-top preferences) and sends to service discovery as the group's preferences to find a proper service. Consequently, the module finds and delivers the best possible service (restaurant) to the group. Meanwhile, from the time that service is delivered at $t_1$ to be used (starting point), to the time the service is used by users, if the relevant context of entities change at $t_2$, which causes changes on users' preferences (e.g., the availability of restaurant changes from open to close, or light traffic changes to heavy traffic in road that users are not willing to take too much time to get to the recommended restaurant), *Context Change Detector* discovers that a change has happened and reports to the discovery module to find a proper service for the current time (e.g., $t_2$, see Fig. 2.). That said, all the entities' profiles will be updated after each change. Thus, the module will provide a new service to meet the users' current preferences, if the new service has higher similarity value compared to the last service w.r.t group preference. This procedure helps to the framework to

**Fig. 2** Satisfaction degree in different situation of context: Blue line shows the satisfaction degree when we run a regular context-aware recommendation, while the green line indicates the output of our proposed recommender system

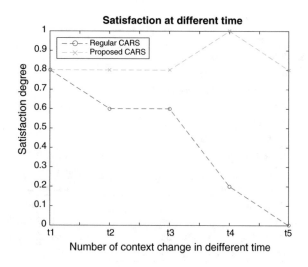

provide the best service at different situations which leads to keep the users satisfaction's degree at high level. As a preliminary implementation, we simulate the framework using Matlab and C++ which is explained in the next section.

## 6 Simulation

The simulation is carried out using MATLAB and C++ where the proposed method is compared to a regular CARS. The simulation describes the effect of the context change on users' preferences and satisfaction degree. In this experiment, we first model 20 restaurants with 5 different features (or context dimensions $N^{CD} = 5$), as well as using GenCPnet [3] to model and generate $N = 5$ CP-nets as the users' conditional preferences. We then use Majority Rule (see preference aggregation module) to aggregate the CP-nets where we obtained a set of preferences, from the most to the least preferable. We only take into account the most preferable restaurant (top one) as the group preference. In order to find the best restaurant in accordance with the group preference, we convert each feature (context dimension) which is used in CPnets and restaurant into a vector (the feature in users' CP-nets and restaurant are the same). Accordingly, we use *Cosine Similarity* to calculate the similarity between group's preferences and all the available restaurants in Space at each time, and then try to find the best restaurant which has high similarity value. Table 1 shows the parameters which are used in this simulation.

**Starting point**; At $t_1$ the system recommends the specific restaurant (which has the nearest similarity (highest value) with respect to the group's preferences) to the group as the best restaurant. And then we randomly change one context dimension (feature) of the recommended restaurant at $t_2$ (the point of the entity "restaurant" is changed to another point "Definition 3"), thereafter, we calculate the satisfaction

**Table 1** Parameters

| Parameters | Value |
|---|---|
| Numbers of users | 5 |
| Numbers of CP-nets | 5 |
| Number of available restaurant | 20 |
| Aggregation rules | Sequential majority |
| Numbers of restaurant features | 5 |
| Domain of each feature | 3 |
| Satisfaction degree | Between 0 (minimum) and 1 (maximum) |
| Features | Value |
| Type of restaurant | 1. italiano, 2. chinese, 3. mexican |
| Quality of the restaurant | 1. low, 2. medium, 3. high |
| Price | 1. cheap, 2. medium, 3. expensive |
| Status of queue at each time | 1. short, 2. normal, 3. long |
| Traffic status around the restaurant | 1. light, 2. medium, 3. heavy |

degree of the group (we normalized the similarity value between 0 to 1 which are the worst and the best (full match) satisfaction degree respectively.). At $t_3$ we change 2 context dimensions of the recommended restaurant, and then again repeating the same procedure to calculate the satisfaction degree. This procedure will continue until all the context[2] are randomly changed. The satisfaction degree in each time is shown with blue line.

**Running the proposed method**; At $t_1$ the RS provides a proper service based on the group's preferences (top one) and calculates the satisfaction degree. At $t_2$ we randomly change 1 context of the recommended restaurant. Thereafter, *Context Change Detector*, detects that a change has happened and reports to the discovery module to find a proper service at $t_2$ based on the current context of the entities. Since the current features of the recommended restaurant might not meet the group's preferences, the satisfaction degree will be evaluated with the group's preferences. Therefore, the new service (restaurant) will be replaced with the previous one (recommended restaurant at $t_1$) at $t_2$, if it has higher satisfaction degree upon the group's preferences. This process will continue until a threshold (e.g., certain time) or the service is used by the group. The concept of different times $(t_1, \ldots, t_5)$ in the simulation is to show the duration between providing a service by the framework and the time that the service might be used by users. We increase the number of changed context at each time and calculate the satisfaction degree which are shown with the green line in the Fig. 2. Y-axis shows the satisfaction degree, and X-axis indicates different times. Although, the two methods start with the same satisfaction level 0.80, at $t_1$, recommendation of our method has higher satisfaction

---

[2]In this experiment we consider that the type of the restaurant will not change at different times, while quality, price, the status of queue (or waiting time) and the status of the traffic around the restaurant will change time to time.

**Table 2** Time, context changes and satisfaction degree

| Time | $t_1$ | | $t_2$ | $t_3$ | $t_4$ | $t_5$ |
|---|---|---|---|---|---|---|
| Number of context change | Initial context/no change | | 1 | 2 | 3 | 4 |
| Satisfaction degree in regular RS | 0.80 | | 0.60 | 0.60 | 0.20 | 0 |
| Satisfaction degree in Proposed RS | 0.80 | | 0.80 | 0.80 | 1 | 0.80 |

degree when there is a context change, even with high number of changes. It needs to be mentioned that at some points the change of context leads to the improvement of satisfaction degree. Having said that, on average, the more the context changes, the more satisfaction degree drops. Table 2 indicates the experiment result, as well as is shown by Fig. 2.

# 7 Conclusion and Future Work

The ultimate aim of the present work was to design a context-aware framework in which services are recommended based on the users' conditional preferences. Preferences are modeled by CP-nets and represented via HAC to handle the dependencies between users' preferences and the complexity of multi-dimensional entities' context, respectively. Sequential Majority rule is used to aggregate users' current preferences (profiles) into the group profile. Thereafter, the recommendation module finds the best possible service that can fulfill the users' requests. Meanwhile, the system can provide new service(s) to the users, if their (one or more) most relevant context of entities changes by affecting the preferences of the group of users. The simulation of the framework showed that the change of the context dramatically decreases the user satisfaction degree, while the proposed framework can keep users at a high level of satisfaction, leading to meet the users conditional preferences. Clearly, further studies are needed to implement the framework with real data, as well as more in-depth investigation into the group's conditional preferences, by considering the importance of context dimensions and priorities among users in particular domain. For example, *price* could be important for a user and *time* could be important for other users in a group.

# References

1. Adomavicius, G., Sankaranarayanan, R., Sen, S., Tuzhilin, A.: Incorporating contextual information in recommender systems using a multidimensional approach. ACM Trans. Inf. Syst. (TOIS) **23**(1), 103–145 (2005)
2. Adomavicius, G., Tuzhilin, A.: Context-aware recommender systems. Recommender Systems Handbook, pp. 217–253. Springer, US (2011)

3. Allen, T.E., Goldsmith, J., Justice, H.E., Mattei, N., Raines, K.: Generating cp-nets uniformly at random. In: Proceedings of the 30th AAAI Conference on Artificial Intelligence (AAAI) (2016)
4. Anand, S.S., Mobasher, B.:. Contextual Recommendation. Springer (2006)
5. Baltrunas, L., Amatriain, X.: Towards time-dependant recommendation based on implicit feedback. In: Workshop on Context-aware Recommender Systems (CARS'09) (2009)
6. Boutilier, C., Brafman, R.I., Domshlak, C., Hoos, H.H., Poole, D.: Cp-nets: A tool for representing and reasoning with conditional ceteris paribus preference statements. J. Artif. Intell. Res. (JAIR) **21**, 135–191 (2004)
7. De Gemmis, M., Iaquinta, L., Lops, P., Musto, C., Narducci, F., Semeraro, G.: Preference learning in recommender systems. Prefer. Learn. **41** (2009)
8. Dey, A.K.: Understanding and using context. Pers. Ubiquit. Comput. **5.1**(2001), 4–7 (2001)
9. Dourish, P.: What we talk about when we talk about context. Pers. Ubiquit. Comput. **8**(1), 19–30 (2004)
10. Gärdenfors, P.: Conceptual Spaces: The geometry of Thought. MIT press (2004)
11. Hothorn, T., Hornik, K., Zeileis, A.: Ctree: Conditional Inference Trees
12. Lang, J.: Graphical representation of ordinal preferences: languages and applications. In: Conceptual Structures: From Information to Intelligence, pp. 3–9. Springer (2010)
13. Lang, J., Xia, L.: Sequential composition of voting rules in multi-issue domains. Math. soc. sci. **57**(3), 304–324 (2009)
14. Li, Q., Kim, B.M.: An approach for combining content-based and collaborative filters. In: Proceedings of the Sixth International Workshop on Information Retrieval with Asian Languages, vol. 11, pp. 17–24. Association for Computational Linguistics (2003)
15. Lu, J., Wu, D., Mao, M., Wang, W., Zhang, G.: Recommender system application developments: a survey. Decis. Support Syst. **74**, 12–32 (2015)
16. Lund, K., Burgess, C.: Producing high-dimensional semantic spaces from lexical co-occurrence. Behav. Res. Methods Instrum. Comput. **1996**, 203–208 (1996)
17. Oku, K., et al.: A recommendation system considering users past/current/future contexts. In: Proceedings of CARS (2010)
18. Palmisano, C., Tuzhilin, A., Gorgoglione, M.: Using context to improve predictive modeling of customers in personalization applications. In: IEEE Transactions on Knowledge and Data Engineering, vol. 20, no. 11, pp. 1535–1549 (2008)
19. Qian, G., Sural, S., Gu, Y., Pramanik, S.: Similarity between euclidean and cosine angle distance for nearest neighbor queries. In: Proceedings of the 2004 ACM Symposium on Applied Computing, pp. 1232–1237. ACM (2004)
20. Rasch, K., Li, F., Sehic, S., Ayani, R., Dustdar, S.: Context-driven personalized service discovery in pervasive environments. World Wide Web **14**(4), 295–319 (2011)
21. Rasch, K., Li, F., Sehic, S., Ayani, R., Dustdar, S.: Automatic description of context-altering services through observational learning. In: Pervasive Computing, pp. 461–477. Springer (2012)
22. Rossi, F.: Preference Reasoning. Logic Programming, pp. 5–8. Springer, Berlin (2005)
23. Rossi, F., Venable, K.B., Walsh, T.: mCP nets: representing and reasoning with preferences of multiple agents. AAAI **4**, 729–734 (2004)
24. Shin, D., Lee, J.-W., Yeon, J., Lee, S.-G.: Context-aware recommendation by aggregating user context. In: IEEE Conference on Commerce and Enterprise Computing, 2009. CEC'09, pp. 423–430. IEEE (2009)
25. Smaaberg, S.F., Shabib, N., Krogstie, J.: A user-study on context-aware group recommendation for concerts. In: HT (Doctoral Consortium/Late-breaking Results/Workshops) (2014)

# Design of Decentralized Inter-Cell Interference Coordination Scheme in LTE Downlink System

**Yen-Wen Chen and Kang-Hao Lo**

**Abstract** Inter-cell interference coordination (ICIC) has become a critical issue in LTE/LTE-A communication system toward efficient radio resource utilization. The basic concept of ICIC is the negotiation of spectrum usage among base stations to prevent interference in cell edge so as to maximize the transmission throughput. In order to achieve effective ICIC, several researches proposed the decision algorithms in centralized manner. However, it is not easy to be implemented in current LTE infrastructure due to the architecture overhead and decision timing constraint. In this paper, we propose a decentralized ICIC algorithms, named relative throughput based resource block coordination (RTRBC) and RTRBC with residual RB (RTRBC_r), to deal with the coordination among eNBs within the existing LTE framework. Through the negotiation for the relative gain and loss in heuristic manner, the proposed scheme negotiates the RB usage between adjacent eNBs to achieve higher throughput. The simulation results show that the proposed scheme performs better than the fractional frequency reuse (FFR) scheme. Additionally, the results indicate that, comparing to the RTRBC scheme, the RTRBC_r scheme can further coordinate the interfered RBs effectively especially when the loads of adjacent cells are unbalanced.

**Keywords** LTE · Inter-cell interference coordination (ICIC) · Resource blocks · Throughput

## 1 Introduction

In addition to the force-majeure factors, such as channel condition or obstacles, the inter cell interference (ICI) is one of the most severe factors that downgrade the transmission quality and system throughput. To get over the ICI problem has

Y.-W. Chen (✉) · K.-H. Lo
Department of Communication Engineering, National Central University,
Tao-Yuan, Jhong-Li, Taiwan
e-mail: ywchen@ce.ncu.edu.tw

© Springer International Publishing Switzerland 2016
R. Lee (ed.), *Software Engineering, Artificial Intelligence, Networking and Parallel/Distributed Computing*, Studies in Computational Intelligence 653, DOI 10.1007/978-3-319-33810-1_9

become one of the most important issues toward effective transmission in cellular networks. Generally, there are two architectures to deal with ICIC. One is the centralized approach and the other is the decentralized approach. Both of them have the same objective, i.e. the achievement of Self-Organized Network (SON). SON can be further divided into three main categories: self-configuration, self-optimization and self-healing [1]. Self-optimization, which ICIC belongs to, includes optimization of coverage, capacity, handover and interference [2]. And the SON architecture can be a centralized, distributed or a hybrid solution.

In centralized ICIC method, the Remote Radio Head (RRH) was proposed as hierarchical flex-grid, and each eNB is a Base Band processing signal Unit (BBU) to manage inter-site Radio Frequency Units (RFUs) through X2 interface [3]. And the Radio over Fiber (RoF) technique is adopted in X2 for better transmission performance in the management plane. Although it successfully turned inter cell coordination into inter sector coordination and having BBU acts as an arbitrator in conflict RBs, this method is not practical enough. In the real situation, it is hard to have the backhaul with RRH model and high computational ability of every BBU.

In [4], the author proposed an ICIC method that based on Harmony Search (HS). According to results of HS, which is obtained by the average interference weight, eNB will execute decentralized arbitration to mute some RBs if they lose in the arbitration. That paper reduced the power consumption and ensured throughput. Moreover, it offered a decentralized coordination method and minimized inter-cell message exchange. However, the improvement of throughput is not significant and the complexity is too high. In [5, 6], the authors took the UEs' GPS information into consideration to enhance the accuracy of estimating main interferer. And they also proposed a simple way to approximately check interferer cell by checking Reference Signal Received Power (RSRP) and derived an equation to shows how much power one eNB should reduce. Nevertheless, it is not practical enough since eNBs cannot always acquire UEs' GPS information. Additionally, to reduce eNBs power on certain RB is hard to be implemented because of the restriction of RF equipment and Reference Signal (RS) issue. In [7], the author suggested to assign conflicted RBs by comparing the traffic load of two adjacent cells. The basic concept is to let eNB with heavier traffic load have higher proportion to obtain the conflicted RB. However, the impact of channel condition was not well considered. And if two eNBs have similar traffic load, it will compare their identifier of cells, which means that certain eNB would always win the arbitration.

The rest of this paper is organized as follows. The system model is provided in the following section. The proposed decentralized RTRBC based scheme is described in Sect. 3. The simulation results are illustrated with comparisons to study the performance of the proposed schemes in Sect. 4. And, the final section concludes our works.

## 2 System Model

In LTE-A system, eNBs can exchange information through X2 interface for management purpose. This mechanism provides the negotiation among eNBs for radio resource utilization. Specifically, eNB can send/receive Relative Narrowband Tx Power (RNTP), High-Interference Indicator (HII), and Overload Indicator (OI) to/from its neighbor eNBs as reference when performing resource allocation as well as handover 0. The definition of the RNTP indication of the specific physical resource block $(n_{PRB})$, $RNTP(n_{PRB})$, which is the main reference for the decisions of handover and ICIC, is defined as follows:

$$RNTP(n_{PRB}) = \begin{cases} 0 & if \ \frac{E_A(n_{PRB})}{E_{max\_nom}^{(P)}} \leq RNTP_{threshold} \\ 1 & else \end{cases} \tag{1}$$

where $E_A(n_{PRB})$ represents the maximum intended EPRE of UE-specific Physical Downlink Shared Channel (PDSCH) Resource Elements (REs) in OFDM symbols, which does not contain the Reference Signal (RS), in this physical resource block on antenna port $p$ in the considered future time interval. $n_{PRB}$ is the physical resource block number which $n_{PRB} = 0, \ldots, N_{RB}^{DL} - 1$. And according to the specifications [1, 2],

$$RNTP_{threshold} \in \{-\infty, -11, -10, \ldots, +3\}[dB] \tag{2}$$

$$E_{max\_nom}^{(p)} = \frac{P_{max}^{(p)} \frac{1}{\Delta f}}{N_{RB}^{DL} * B_{SC}^{RB}} \tag{3}$$

where $P_{max}^{(p)}$ is the base station maximum output power, $\Delta f$ is the subcarrier spacing, $N_{RB}^{DL}$ is the downlink bandwidth configuration, which expresses in multiples of $N_{SC}^{RB}$, which is the resource block size in the frequency domain [9].

When the ratio of $E_A(n_{PRB})/E_{max\_nom}^{(p)}$ is higher than $RNTP_{threshold}$ in a cell, its' associated eNB shall arise the transmission power on certain PRBs because of the strong interference from adjacent cell(s). And the eNB will send a RNTP message to its neighboring cells to trigger ICIC. In our method, both RNTP and the proposed corresponding RB's ICIC weight $W_i^j$, which denotes the utilization value for the $i$th cell using the $j$th RB, is sent to the neighboring eNBs as well. And, after receiving $W_i^j$ from other eNBs, the eNB shall execute the competition process to determine whether to mute the transmitting power of the $j$th RB. The objective of the proposed method is to maximize the total system throughput but not always favor which RB has better signal quality. Let $f_j$ be the objective function of RB $j$, then

$$f_j = Max \left[ \sum_{i=1}^{I} \sum_{\forall u \in U_i} R_{u,i}^j \right] \tag{4}$$

where $I$ is the total number of cell, $U_i$ denotes the set of UEs in cell $i$, and $R_{u,i}^j$ represents the data rate of UE $u$ of cell $i$ in RB $j$. Figure 1 illustrates the proposed operational model. There are 3 stages of the proposed scheme. Stage 1, which follows the 3gpp specification, determines whether there is trigger indication for ICIC request according to Eq. (1). Stage 2 and 3 are the main part of the proposed RTRBC scheme, in which, stage 2 determines the weighting function and stage 3 performs the virtual competition and RB allocation. These two stages will be discussed in the following section.

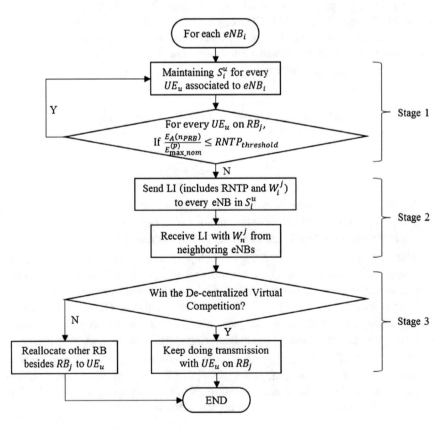

**Fig. 1** The proposed ICIC operation model

## 3   The Proposed Distributed ICIC Scheme

(A)   Weighting Functions Design

The proposed RTRBC scheme adopts the weighting function, which provides the concept of willing-to-pay of RB bandwidth, for RB coordination in distributed manner. Basically, the weighting value is calculated by each eNB and sent with RNTP in LI through X2 interface to neighboring cells for competition. The weighting function considers the relative achievable throughputs of neighbor eNBs for the competitive RB(s). The basic weighting function of RTRBC is designed as

$$W_{RTRBC\,i}{}^{j} = \frac{R_{u,i}^{j(ach)}}{\bar{C}_{u,i}} \tag{5}$$

where the numerator $R_{u,i}^{j(ach)}$ is the achievable data rate of UE $u$ of cell $i$ in RB $j$ without any interference, and the denominator $\bar{C}_{u,i}$ is the achievable average data rate of UE $u$ in cell $i$ when applying the residual RB. Thus,

$$R_{u,i}^{j(ach)} = D\left(\frac{SINR_{u,i}^{j} \cdot (L\_RSRP_{u,i}^{j} + N_0)}{N_0}\right) \tag{6}$$

and

$$\bar{C}_{u,i} = \frac{\sum_{j \in S_{r\_RB}^{i}} D(SINR_{u,i}^{j})}{N_{r\_RB}^{i}} \tag{7}$$

where $SINR_{u,i}^{j}$ is the SINR value of UE $u$ of cell $i$ in RB $j$, $L\_RSRP_{u,i}^{j}$ represents the largest RSRP value among the adjacent cells reported by UE $u$ of cell $i$ in RB $j$, $N_0$ is the thermal noise power density [10]. $D(.)$ is the adaptive modulation and coding (AMC) function that maps the SINR value to the associated data rate. And $S_{r\_RB}^{i}$ and $N_{r\_RB}^{i}$ denotes the set and the number of residual RB in cell $i$, respectively. It is noted that Eq. (5) provides the throughput comparison of RB $j$ and the average of residual RBs, it reflects the preference of UE $u$ to keep RB $j$ in cell $i$. Then the higher value of the weighting means the more preference of that RB. If eNB $i1$ with smaller weighting value of a specific RB gives up and mute it to let the other eNB $i2$ exclusively use that RB, we pre-assume that eNB $i1$ shall be able to allocate one of the other RB to the UE for compensation. However, as eNBs exchange LI messages in batch and the RB allocation is performed in one by one manner, the UE, who gives up the specific RB owing to its smaller weighting value, may be allocated with the RB having poor condition or even no residual RB for it. Generally, eNB with more free RBs for allocation has more chance to allocate proper RB(s) to compensate its UE. It leads to the heuristic that eNB, which has more available

RBs, tends to conservative to contend for the interfered RB. Then we put a factor $w_i$ into the weighting function, $W_{RTRBC\_r i}{}^j$, when the number of available RB is considered as shown in the following equation.

$$W_{RTRBC\_r i}{}^j = w_i \cdot \frac{R_{u,i}^{j(ach)}}{\overline{C}_{u,i}} \tag{8}$$

Basically the value of $w_i$ decreases as the number of available RB increases. And its impact is more significant when the number of residual RB is quite limited. For example, if there are five RBs need to be negotiated by eNB A and eNB B, and eNB A and eNB B win 1 RB and 4 RBs, respectively, after comparing their $W_{RTRBC}$ values. However, the numbers of available RBs in eNB A and eNB B are 6 and 0, respectively. In this case, eNB A still have enough RBs to be flexibly chosen though it loses 4 RB in the competition. On the contrary, eNB B cannot compensate its muted RB with any bandwidth because there is no residual RB. Thus, we design $w_i(N_{r\_RB}^i)$ to be inverse proportional to the number of the remaining available RB as

$$w_i(N_{r\_RB}^i) = \left( \frac{e^{-f(N_{r\_RB}^i)}}{2} + 0.5 \right) \tag{9}$$

It is clear that $1 \geq w_i > 0.5$ in Eq. (9) if the function $f(N_{r\_RB}^i)$ follows the following boundary conditions:

$$f(N_{r\_RB}^i) = \begin{cases} 0 & \text{if } N_{r\_RB}^i = 0 \\ \infty & \text{if } N_{r\_RB}^i = N_{RB} \end{cases} \tag{10}$$

$N_{RB}$ denotes the maximum number of RB according to the transmission bandwidth configuration (e.g. $N_{RB} = 100$ if the channel bandwidth is 20 MHz).

(B)  De-centralized Virtual Competition and RB Re-allocation

Once the eNB receives the weighting functions together with RNTP values from its adjacent eNBs through X2 interfaces, the de-centralized virtual competition procedure is performed to decide which eNB shall mute the conflicted RB. Thus the eNB can still utilize the RB if it has the highest weighting value in that RB when compared to the other neighbor eNBs. Thus let $\varphi_{i,j}$ be the operation of eNB $i$ on RB $j$, $\varphi_{i,j} = 1$ means eNB $i$ can keep using RB $j$ and $\varphi_{i,j} = 0$ denotes the eNB $i$ shall mute this RB. Then

$$\varphi_{i,j} = \begin{cases} 1 & \text{if } W_i^j = \underset{k \in R_{adj}^{i,j}}{Max} [W_k^j] \\ 0 & \text{otherwise} \end{cases} \tag{11}$$

where $W_k^j$ can be either $W_{RTRBC\,k}{}^j$ or $W_{RTRBC\_r\,k}{}^j$ depending on which weighting function is applied. $R_{adj}^{i,j}$ is the set of the neighbors of eNB $i$ that have sent the weighting function associated with RB $j$ to eNB $i$.

The other eNB shall mute its RB that does not have the highest weighting value and reallocate RB to its UE from its residual RBs. It is possible that the UE may be allocated with the RB which has poor channel condition than its original RB. And the UE may not be allocated with any RB if there is no RB left.

# 4 Experimental Simulations

In order to examine the performance of the proposed distributed ICIC schemes exhaustive simulations were carried out. The simulation model consists of seven eNBs and several UEs using OFDMA for downlink transmission. We further assume the channel bandwidth to be 10 MHz. And there are 50 RBs in the system per TTI per cell. There are 28 levels of MCS in the simulation model. In our simulation, each cycle sustains 1000 TTIs in 1 s. The cycle of CQI report is assumed to be 10 ms and the ICIC procedure is performed every 20 ms according to the information exchange period of X2 interface [11]. The type 1 FDD is utilized and the channel condition is changeable but is assumed to be stable within one second. The parameters used for simulations are listed in Table 1.

The $f(N_{r\_RB}^i)$ used in Eq. (9) was assumed to be $0.075 * (100N_{r\_RB}^i)/N_{RB}$ to satisfy Eq. (10). Hence the following $w_i(N_{r\_RB}^i)$ was applied during simulations.

$$w_i(N_{r\_RB}^i) = \left( \frac{e^{-0.075\frac{100}{N_{RB}}N_{r\_RB}^i}}{2} + 0.5 \right)$$

**Table 1** Simulation Parameters

| Parameters | Values |
| --- | --- |
| Scenario | 7 cells with 7 eNBs |
| eNB Inter-Site distance (ISD) | 500 m |
| Channel bandwidth | 10 MHz |
| Total resource block | 50 RBs |
| Total bandwidth per RB | 180 kHz |
| Pathloss model | L = 128.1 + 37.6 $log(R)dB$, R[km] [12] |
| Traffic scenario | Full buffer model |
| Total eNB TX power | 46 dBm |
| BS antenna gain | 14 dBi |
| UE noise figure | 9 dB |

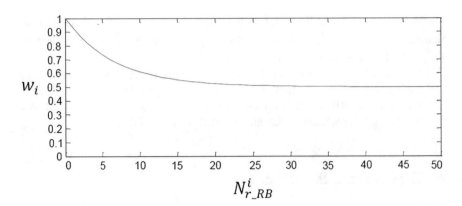

**Fig. 2** The change of $w_i$ versus $N^i_{r\_RB}$

Figure 2 illustrates the $w_i(N^i_{r\_RB})$ values versus the number of residual RB when $N_{RB} = 50$ (i.e. channel bandwidth is 10 MHz). It shows that the designed factor $w_i$ has significant effect for the number of residual RBs is <15.

The average throughput for all UEs and cell edge UEs were examined by changing the percentage of UEs in the cell edge. As the channel bandwidth was assumed to be 10 MHz, there are 50 RBs in each eNB. The results of the proposed RTRBC and RTRBC_r schemes are compared to the results without ICIC and Fractional Frequency Reuse (FFR).

- Case 1: Uniform Traffic Load

In the uniform distribution case, every cell has the same resource saturation ratio and the percentage of UEs. We assume that each UE utilized one RB and there were 280 UEs. Thus the resource saturation ratio was 0.8, i.e. 280/(50*7), during the simulations. The main objective of this case is to investigate the change of throughput with respect to the ratios of cell edge UE in the cell edge for different saturation ratios. Figures 3 and 4 show the average throughput of all UEs and UEs in the cell edge, respectively. It is clear that the difference of overall throughputs are not significant, however, the proposed schemes, either RTRBC or RTRBC_r, achieves much better throughput than the other two schemes when only UEs in the cell edge are considered, especially when the percentage of cell edge UEs is higher.

- Case 2: Unbalanced Traffic Load

In this case, we analyze the negotiation of RBs between two eNBs with unbalance saturation ratios. The two-cell model for simulation is given in the following Fig. 5. Thus, the hotspot cell fixed its saturation ratio to be 0.9 and its neighbor cell varied the saturation ratio from 0.18 to 0.9.

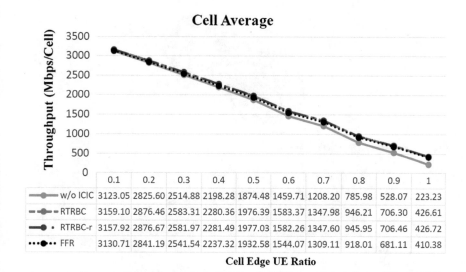

**Fig. 3** Overall average throughput versus cell edge UE ratios

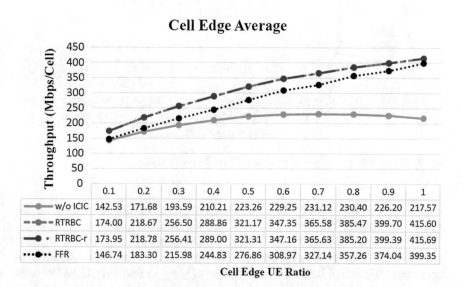

**Fig. 4** Average throughput of cell edge UEs versus cell edge UE ratios

The average throughput of cell edge UEs is provided in Fig. 6. It illustrates that the RTRBC_r scheme achieves higher throughput than the RTRBC scheme. We further examine the cell edge throughput of the hotspot cell. As shown in Fig. 7, the

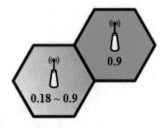

**Fig. 5** the Two-cell simulation model

## Two Cell Edge Average

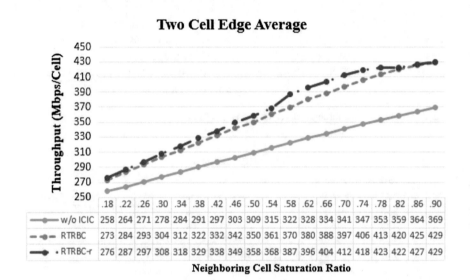

| | .18 | .22 | .26 | .30 | .34 | .38 | .42 | .46 | .50 | .54 | .58 | .62 | .66 | .70 | .74 | .78 | .82 | .86 | .90 |
|---|---|---|---|---|---|---|---|---|---|---|---|---|---|---|---|---|---|---|---|
| w/o ICIC | 258 | 264 | 271 | 278 | 284 | 291 | 297 | 303 | 309 | 315 | 322 | 328 | 334 | 341 | 347 | 353 | 359 | 364 | 369 |
| RTRBC | 273 | 284 | 293 | 304 | 312 | 322 | 332 | 342 | 350 | 361 | 370 | 380 | 388 | 397 | 406 | 413 | 420 | 425 | 429 |
| RTRBC-r | 276 | 287 | 297 | 308 | 318 | 329 | 338 | 349 | 358 | 368 | 387 | 396 | 404 | 412 | 418 | 423 | 422 | 427 | 429 |

**Neighboring Cell Saturation Ratio**

**Fig. 6** The average throughput of cell edge UEs in Two-cell model

UE in the hotspot cell edge receives higher average throughput when applying the RTRBC_r scheme. The main reason is that the weighting function $W_{RTRBC\_r}{}_i{}^j$ considers the residual number of RBs when performing RB negotiation. The hotspot cell tends to keep using the interfered RB and it also pushes its neighbor cell to mute the interfered RB because of the smaller $w_i(N^i_{r\_RB})$ value, even if the UE of its neighbor cell has higher achievable data rate on that RB. Otherwise, if the hotspot cell makes the concession, it may not be able to allocate RB to its UE due to the depletion of RB. Although the neighbor eNB gives way to the hotspot eNB on some RBs, it has more chance to reallocate its residual RB to compensate its UE. And this is the purpose of applying the factor $w_i(N^i_{r\_RB})$.

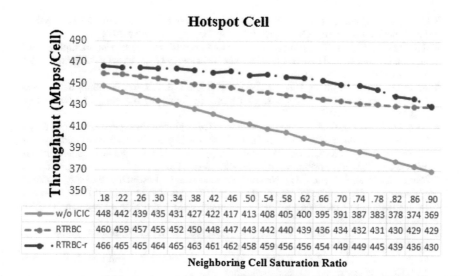

| | .18 | .22 | .26 | .30 | .34 | .38 | .42 | .46 | .50 | .54 | .58 | .62 | .66 | .70 | .74 | .78 | .82 | .86 | .90 |
|---|---|---|---|---|---|---|---|---|---|---|---|---|---|---|---|---|---|---|---|
| w/o ICIC | 448 | 442 | 439 | 435 | 431 | 427 | 422 | 417 | 413 | 408 | 405 | 400 | 395 | 391 | 387 | 383 | 378 | 374 | 369 |
| RTRBC | 460 | 459 | 457 | 455 | 452 | 450 | 448 | 447 | 443 | 442 | 440 | 439 | 436 | 434 | 432 | 431 | 430 | 429 | 429 |
| RTRBC-r | 466 | 465 | 465 | 464 | 465 | 463 | 461 | 462 | 458 | 459 | 456 | 456 | 454 | 449 | 449 | 445 | 439 | 436 | 430 |

**Neighboring Cell Saturation Ratio**

**Fig. 7** The average throughput of cell edge UEs in the Hotspot cell

## 5  Conclusions

In this paper, the relative throughput based RB coordination scheme is proposed to increase the throughput in LTE downlink system. By considering the achievable rates of interfered RB and the potential alternative RB, the proposed RTRBC scheme can achieve higher throughput when comparing to the other scheme. Moreover, the proposed RTRBC_r scheme, which takes the remaining number of RB into consideration, can further effectively coordinate the utilization of interfered RB especially when the loads of adjacent eNBs are unbalanced. The simulation results demonstrate that the proposed schemes are superior to the other schemes. The proposed schemes perform the coordination in decentralized manner under the LTE architecture. Hence it is applicable for current LTE operations.

**Acknowledgement** This research work was supported in part by the grants from the National Science Council (NSC) and Ministry of Science and Technology (MOST) (grant numbers: 102-2221-E-008-015, 103-2221_E008-082, 104-2221-E-008-050, and 104-3115-E-194-001), Taiwan, ROC.

## References

1. 3GPP TS 32.500 (v12.1.0 Release 12): Telecommunication management; Self-Organizing Networks (SON); Concepts and requirements, Dec 2014
2. 3GPP TS 32.511 (v12.0.0 Release 12): Telecommunication management; Automatic Neighbour Relation (ANR) management; Concepts and requirements, Oct 2014

3. Ogata, D., Nagate, A., Fujii, T.: Multi-BS cooperative interference control for LTE systems. In: Vehicular Technology Conference (VTC Spring), pp. 1–5, May 2012
4. S.S., Khalifa, Hamza, H.S., Elsayed, K.: Inter-Cell interference coordination for highly mobile users in LTE-Advanced systems. In: Vehicular Technology Conference (VTC Spring), pp. 1–5. Cairo University, June 2013
5. Xiao, D.: Beijing Institute, Huawei Technologies Co. Ltd. A novel downlink ICIC method based on user position in LTE-advanced systems. In: Vehicular Technology Conference (vtc fall), pp. 1–5. Sept 2012
6. Xiao, D.: Beijing Institute, Huawei Technologies Co., Ltd. Jian Huang & Xiaojun Jing Beijing University. A downlink ICIC method based on region in LTE-Advanced systems. International Symposium on Personal, Indoor and Mobile Radio Communications Workshops (PIMRC Workshops), pp. 420–423. Sept 2010
7. Wang, S., Zhang, Y., Bi, G.: National Communication Research Laboratory Southeast University. A decentralized downlink dynamic ICIC method for Multi-cell OFDMA system. In: International Conference on Wireless Communications and Signal Processing (WCSP), pp. 1–5, Nov 2011
8. 3GPP TS 36.213 (v12.5.0 Release 12): Evolved Universal Terrestrial Radio Access (E-UTRA); Physical layer procedures, Mar 2015
9. 3GPP TS 36.211 (v12.5.0 Release 12): Evolved Universal Terrestrial Radio Access (E-UTRA); Physical channels and modulation, Mar 2015
10. 3GPP TS 36.133 (v12.5.0 Release 12): Evolved Universal Terrestrial Radio Access (E-UTRA); Requirements for support of radio resource management, Oct 2014
11. 3GPP TS 36.423 (v12.5.0 Release 12): Evolved Universal Terrestrial Radio Access Network (E-UTRAN); X2 Application Protocol (X2AP), Mar 2015
12. 3GPP TR 36.942 (v12.0.0 Release 12): Evolved Universal Terrestrial Radio Access (E-UTRA); Radio Frequency (RF) system scenarios, Oct 2014

# Toward Flow-Based Ontology

**Sabah Al-Fedaghi**

**Abstract** Now, as we progress in this age of information and communication technology, various domains of science and life must face the problem of overwhelming amounts of data being made available; the aim here is to improve retrieval and dissemination of information. One solution is to build *ontologies* with fundamental roles to support knowledge sharing and reuse. An ontology for a certain domain provides a uniform representation of information, thus facilitating interoperability and sharing of that information. Most current approaches to ontology try to refine relationships among basic categories in a hierarchical topology through such multirelationships as *located_in*, *adjacent_to*, *transformation_of*, *preceded_by*, *has_agent*, and so on. This paper introduces an exploratory study of these multirelationships with the aim of adding a more systematic foundation that visualizes structure and incorporates procedures in the form of input-process-output. The paper proposes that the notion of *flow* can serve Zas a tool to support the hierarchical assembly of terms comprising the ontology. The flow-based representation is demonstrated with examples from the known Biomedical Ethics Ontology (BMEO) and others. The resultant contrast seems to indicate a viable technique that can be incorporated into building ontologies for other domains.

**Keywords** Conceptual model · Ontology · Domain ontology · Knowledge representation

## 1 Introduction

Now, as we progress in this age of information and communication technology, various domains of science and life are making use of the World Wide Web and of locally held databases to interact and exchange information. Accordingly, we have

S. Al-Fedaghi (✉)
Computer Engineering Department, Kuwait University, P.O. Box 5969,
13060 Safat, Kuwait
e-mail: sabah@alfedaghi.com

© Springer International Publishing Switzerland 2016
R. Lee (ed.), *Software Engineering, Artificial Intelligence, Networking and Parallel/Distributed Computing*, Studies in Computational Intelligence 653, DOI 10.1007/978-3-319-33810-1_10

become overwhelmed by the vast amount of data being made available [1]. The problem is building computerized databases in such a way that they can be reasonably used by persons.

Currently (2010), the terminologies used in conveying and organizing information in distinct domains are developed in ad hoc ways; often, terminologies and database schemas fall short of being interoperable even when prepared by researchers from the same departments, groups, or labs… The result is a silo effect: data and information are isolated in multiple, incompatible silos, and shareability and reusability is greatly limited [2].

One aspect of the required solution is building ontologies with the fundamental role of supporting knowledge sharing and reuse [3]. Informally, an ontology comprises concepts and relationships that describe and constrain how the concepts refer, interrelate, and combine [4]. *Traditionally*, the term *ontology* is used in philosophy to refer to the metaphysical study of the world with the aim of classifying the world's entities and relationships. The resultant representation is a taxonomy that is structured as a hierarchy of types, universals, or classes that branch into subtype relationships. Currently, it is proposed that the problem of organizing computerized databases can be alleviated by utilizing *domain ontologies*. A domain is a subject about which we seek to acquire knowledge. A domain ontology is a taxonomic **representation** of the entities and relationships existing within a particular domain of reality [2, 5–7]. An ontology in a certain domain improves the search for information, makes information processing more efficient, and makes data from diverse sources more consistent. Ontology also provides a uniform representation of information, thus facilitating interoperability and sharing of that information. In such ontologies,

The nodes are connected by edges representing principally the is_a subtype relation, but also supplemented by other edges representing binary relations such as *part_of*, *preceded_by*, *has_participant*, *inheres_in*, and other relations holding between these types of entities…

Further, the domain ontology contains properties and axioms that are designed to enable algorithmic reasoning on the basis of these relationships, so that new information about the underlying instances that comprise the domain of study might be inferred [2].

Most current methods of building a domain ontology involve [2]:

- Determining the basic universals and relations,
- Using Aristotelian structure to formulate definitions.
- Putting the universal terms in a taxonomic hierarchy, and adding the relevant relationships.

Current logic tools and formal languages for reasoning systems available for use with ontologies are still fairly limited in their capacities [8, 10]. There is a need for progress in the field of ontology "by pointing out the pitfalls of poor *representation* and reasoning that hamper information accessibility and dissemination" [2; italics added].

Note that this paper focuses on representation, leaving discussion of the reasoning process to future research. The proposition adopted in the paper is that a "good" representation of ontology should be measured in terms of systemization while applying the classical system framework of input-process-output. A systematic representation in this context can be developed. The paper proposes that the notion of flow can serve as a supplementary tool to model relationships in a hierarchical ontology. To demonstrate the viability of this notion, the resultant representation is contrasted with examples from the known Biomedical Ethics Ontology (BMEO) and others.

A model called the Flowthing Model (FM) will be utilized in this scheme. It has been used in several applications (e.g., [9–13]). For the sake of a complete presentation, the basic FM will be briefly described in the next section. The negotiation modeling example, developed there to illustrate the nature of the FM representation, is a new contribution. In Sect. 3, certain aspects of the Aristotelian categories are viewed in terms of FM. Section 4 discusses the known Biomedical Ethics Ontology and recasts some of its multi-relationships in terms of systematic flow. Section 5 strengthens the claim of the descriptive capability of the proposed method by applying it to the classical ontological notion of *definitions*.

## 2   Flowthing Model

FM is a system structured of stages among which things move as well as into and out of the system. Everything participates in such a system or part of one. The model takes the classical form of intersecting and overlapping divisions or spheres (e.g., Venn diagrams, a hierarchical structure of ontology) but these spheres embed *flows* that move across boundaries of intersecting and nested spheres. Components of a flow include:

- flowthings ("things that flow," or artifacts) and
- flow systems (flowsystems) that represent the riverbeds of flows, each with at most six stages (see Fig. 1).

Messages, ideas, concepts, and opinions are examples of flowthings or flowsystems. A thing is defined in FM as a flowthing: that which is created, processed, arrives, is accepted, released, and transferred, while flowing within and among spheres. It has a permanent identity but impermanent form, e.g., the same news translated into different languages.

An abstract flowsystem constrains the trajectory of flow of flowthings. The flowsystem shown in Fig. 1 is a generalization of the traditional input-process-

**Fig. 1** Flow system

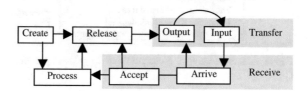

**Fig. 2** Input-process-output model

output model used in many fields (Fig. 2). It comprises a delineation of the distinct parts and operations in the geography of flow of a system.

Flows connect the five stages (transfer, release, process, receive, and create) that are exclusive for flowthings; i.e., a basic flowthing can be in one and only one of these states at a time. A *stage* here is a "transmigration field" of the flowthing that is created, processed, released, and transferred, arrives, and is accepted (or simply received, combining arrived and accepted into one state). In Fig. 1, irreversibility of flow is assumed, e.g., released flowthings flow only to Transfer.

The exclusiveness of FM stages (i.e., a flowthing cannot be in two stages simultaneously) indicates synchronized change of the flowthing, i.e., a flowthing cannot be changed in form and sphere simultaneously. This is a basic systematic property of flowthings.

FM uses the notions of *spheres and subspheres*. These are the network environments and relationships of systems and subsystems. A sphere can be a mind, an organ, an entity (e.g., a company, a customer), a location, a communication medium (a channel). FM also utilizes the notion of triggering. Triggering is the activation of a flow, denoted in FM diagrams by a *dashed arrow*. It is a dependency among flows and parts of flows. A flow is said to be triggered if it is created or activated by another flow (e.g., a flow of electricity triggers a flow of heat), or activated by another point in the flow. Triggering can also be used to initiate events such as starting up a machine (e.g., remote signal to turn on).

*Example* Galitsky et al. [14] developed a knowledge representation methodology based on graphs and a natural language scenario of an inter-human conflict scenario as a sequence of communicative actions, ordered in time, with a causal relation between certain communicative actions.

Scenarios are simplified to allow for effective matching by means of graphs: only communicative actions remain as a most important component to reflect the dialogue structure and express similarities between scenarios. Each vertex corresponds

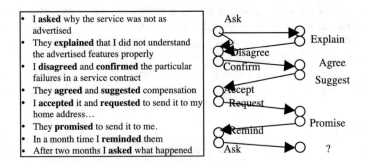

**Fig. 3** A sample complaint scenario and its graph representation (partial, from Galitsky et al. [14])

to a communicative action, which is performed by either proponent, or opponent. An arc (oriented edge) denotes a sequence of two actions [14].

Figure 3 shows an example of a scenario and its graph. It involves inter-human conflict or negotiation between *I* and *They*. Figure 4 shows the corresponding FM representation.

Figure 4 includes different spheres of specified time increments (outer boxes) and flowsystems of questions, confirmations, agreements, etc. In the figure at

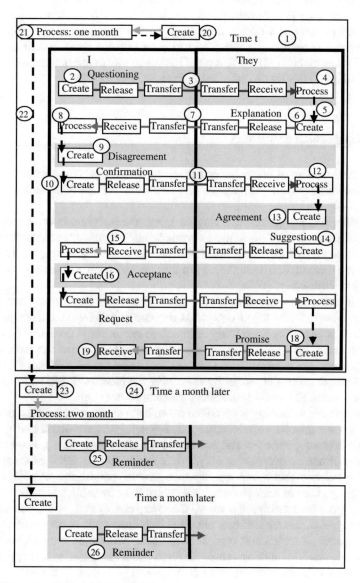

**Fig. 4** FM representation of the scenario

certain time $t$ (circle 1 in the figure), I create (circle 2) questions regarding the service that flow (3) to Them. When They process the questions (4), this triggers (5) creating (6) and sending their explanation (7). When I process their explanation (8), this triggers disagreement (9) about it, causing me to create a confirmation (of their failure—10) that flows to them (11). They process my confirmation (12), leading them to agree with it (13) and suggest (14) compensation. Upon receiving their suggestion (15), I accept (16) and make a request (17) that they act on it. Accordingly, they make a promise (18) and send it to me (19).

Note that time is a *flowthing*; thus, it is created (20—acknowledged in the conceptual picture) at a certain time, $t1$. It is also processed (21), which means being *consumed* until one month later to trigger (22) the creation (23) of another eventful point in time (24). At that point in time (a month later), a reminder is sent (25). Similarly, after two months another reminder is sent (26).

# 3 Aristotelian Ontology

To illustrate the aim of this paper of a flow-based ontology, consider the following discussion. McPartland [15] reports that Aristotle considers *argument through names*:

For since it is not possible to bring the things themselves into conversation, we instead use the names of the things as symbols, and we suppose that the things that follow in the case of the names also follow in the case of the things, just like those performing a calculation do in the case of the counters ([15], referencing Aristotle's *Sophistical Refutations* (165a, 6–10).

In this counting, every time a cow walks into a pen (enclosure), one rock is placed in a bowl. Every time a cow walks out, one rock is removed from that bowl. Provided this procedure is correctly followed, someone could determine how many cows are in the pen by counting the number of rocks in the bowl. An isomorphism exists between the cow-pen world and the rock-bowl model, and this fact allows one to find out about the former by examining the latter [15]. We will model this procedure using FM and a hierarchical structure of ontology.

An Aristotelian ontology is described through the notion of categories, or entities with a known inheritance relationship. The ontology is supplemented by distinguishing between particulars and universals, and the well-known general kinds of entities—substance, quantity, quality, relationship, and so forth. Since this is a computer science paper and not oriented toward philosophy or philosophical knowledge, it tries to limit philosophical controversy by using a minimal basis for ontology (for characterizing the world) at the expense of embedding philosophical inadequacy, e.g., mixing different philosophical terms such as universal, property, quality, feature, characteristic, type, kind, attribute, sort, etc. In addition, we assume intuitions made in mixing cases involve both universals and particulars. A discussion of such issues as *what is "universal"?* (e.g., whether universals are repeatable entities or the nominalist rejection of that), is sacrificed for the sake of the bigger picture of *representation* of the fabric of the ontology (hierarchical structure plus cross-hierarchy relationships).

To conceptualize the fabric (conceptual structure) of the world, Aristotle had to establish relationships among categories, e.g., one thing is predicated of another ("a relation between words [linguistic] and a relation between entities [metaphysical]," Aristotle's version of a correspondence theory of truth [15]). *Conceptualization* here refers to "a set of informal rules that constrain the structure of a piece of reality, which an agent uses in order to isolate and organize relevant objects and relevant relations" [16]. A similar conceptualization is given by Plato, who "tells us that an assertion requires that a verb and a name be *woven together*, and that a simple list of words is not yet an assertion. Plato also seems to think that true speech requires a *weaving together* of entities in the world" [15], referencing [17]. Also, in this fabric, logical, causal, and phenomenal *properties* (as relationships among things) are used to enforce some order of reality.

In this case, logic and truth as a symbolic system mirrored in the "real" system (World) are utilized to weave together categories and their relationships and to build a horizontal fabric in addition to the vertical hierarchical structure of ontology. The truth of linguistic assertion (relationships among terms) is the truth of metaphysical cross-category (horizontal) relationships.

Most current approaches to ontology try to refine relationships among categories, thus completing the cross-category fabric, through such multi-relationships as *located_in, contained_in, adjacent_to, transformation_of, derives_from, preceded_by, has_participant, has_agent,* and so on (see Fig. 5).

Alternatively, the general claim in this paper is that the notion of *flow* can serve as a tool to model relationships in this hierarchy. Thus, for example, a *preceded_by* relationship such as *A cow reaches a farm then it enters a pen in the farm* is represented as *A cow flows to a farm then to a pen in the farm* (see Fig. 6), taking into consideration a future instantiation of some physical state. One advantage is

**Fig. 5** Illustration of structuring of the World in terms of a vertical ontology of categories woven together in diverse relationships

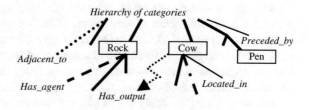

**Fig. 6** A cow in the pen

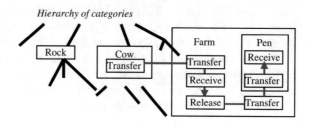

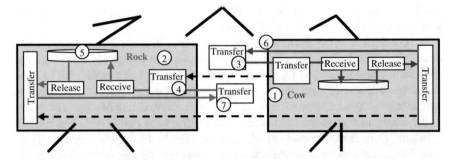

**Fig. 7** Counting cows: each time a cow walks into a pen, one rock is placed in a bowl

that the flow itself and the five stages of Create, Release, Transfer, Receive, and Process encompass many relationships. Accordingly, Fig. 7 shows a representation of Aristotle's cows/rocks analogy. In the context of the hierarchy, a "bridge" is built between Cow (1) and Rock (2). Each time a cow enters a pen (3), it triggers a rock to flow (4) into storage (5).

Each time a cow moves out (6), a rock is triggered to flow out of storage (7); thus *processes* are superimposed on the hierarchical structure. The proposition is that such a mixture of structure and flows seems a promising approach to underlying any reasoning process instead of the rigid addition of edges to the hierarchy of ontology, as will be illustrated next.

## 4 Biomedical Ethics Ontology

According to Koepsell et al. [18], a Biomedical Ethics Ontology (BMEO) is used by ethics committees to deal with protocols and consent (see Fig. 8). These committees face an overabundance of information, changing rules and requirements, as well as a constant stream of new cases. "Ontology would make it possible to archive the results of ethics committee deliberations, search that information intelligently, and put it to future use in a number of ways" [18]. Benefits of BMEO include improved clarity or consensus on applying ethical principles in similar situations, and increased efficiency, accuracy, and standardization [18].

Koepsell et al. [18] apply BMEO in the domain of human subjects research. First, this domain ontology involves generating entities associated with biomedical ethics in the IRB context. They provide "Aristotelian definitions—definitions of the form "'A is_a B that has C' or 'A is_a B that Cs'—for each" [18]. The ontology incorporates other relationships such as *beyond is_a*, and *owes_a* for ethical duties.

We have described only a very small snapshot of a standard ethics committee review process and relations among its objects. Much more must be done before use can be made of what we are building, and contributions from others examining the domain will no doubt be necessary [18].

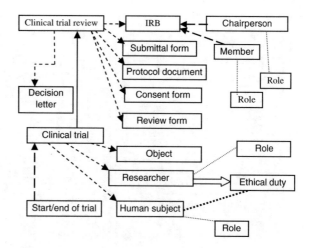

**Fig. 8** Beginnings of a BMEO with a focus on IRBs ([18]; names and labels have been simplified)

Nevertheless, such an approach is not based on a systematic foundation that visualizes structure and incorporates process as described in the example of cows and rocks. In their figure (Fig. 8), *Start/end* of trial is *Part_of* (indicated by the type of edge) Clinical trial. Where is the position of this *Start/end* in the classical framework of input-process-output? It has been scattered everywhere, e.g., *Decision letter* has been labeled *has_output*. The point here is that this multiplicity of relationships in the ontology may be beneficial for enhancing the structure of the hierarchy, but it does not facilitate the incorporation of processes.

To highlight such concerns, the corresponding FM representation is developed (Fig. 9). As an introduction to this representation, Fig. 10 shows it in terms of a hierarchal structure similar to Koepsell et al.'s [18] diagram. The picture is modified and called *Clinical trial case*, showing the flow of a decision made by IRB regarding an application for Clinical trial; thus we avoid ambiguity involving the term "review" (by not including it in title but as an output). In addition, our diagram does not include the *Protocol document* since we show several other documents (forms and fill forms).

Accordingly, the ontology description involves a *Clinical trial case* that connects *IRB* and *Clinical trial* (circles 1, 2, and 3 in Fig. 10, respectively). Consent (4) and Clinical trial data (5) flow to the IRB, where they trigger (6) the writing of a review of the case (7) that triggers creation of a decision that flows to the applicant.

Figure 9 shows the complete FM representation. First, IRB provides a standard form for consent (circle 1); this can be downloaded to flow (2) to the clinical trial, where it is processed (3) to create a filled consent form (4) that is sent to IRB (5). Similarly, a clinical trial form (6) flows to the clinical trial (7) to be processed (8) to trigger the creation (9) of the filled clinical trial form (10). Creating a clinical trial means creating the object (11), a researcher (12), a human subject (13), and the start and end (14) of the trial's ethical duty statement (15).

The ethical duty statement involves the participation of both the researcher and the human subject; thus it is created at the intersection of these two subspheres. The

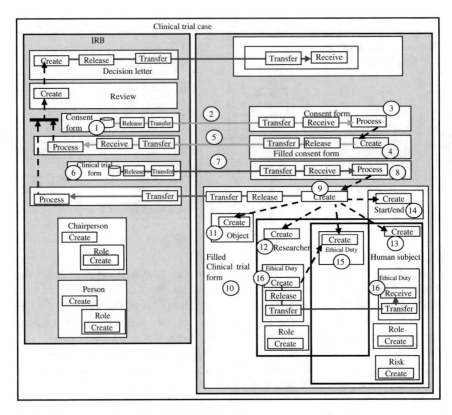

**Fig. 9** FM representation of the IRB/Clinical trial case

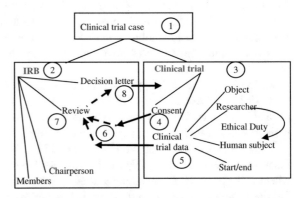

**Fig. 10** Introductory illustration of the FM representation

duty flows *from* the researcher *to* the human subject (16 and 17, respectively). This declaration or creation of duty going in this direction triggers the creation of the ethical duty statement. The representation simply states that because the researcher has an ethical duty to the human subject, a statement is created defining and

**Fig. 11** The researcher-human subject-ethical duty relationship as represented by [18] (*left*) and in FM (*right*)

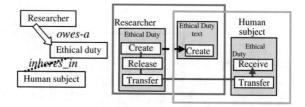

explaining it. This is represented in the ontology of Koepsell et al. [18] (Fig. 8) by the relationship *owes-a* between the researcher and his or her ethical duty (large horizontal arrow), and *inheres_in* between the human subject and ethical duty (see Fig. 11). Note how the FM representation uses the same notions (flow and triggering) to represent the researcher–human subject–ethical duty relationship. No additional graphic symbols or linguistic labels are necessary.

## 5  Definitions

A current paradigm in domain ontology is the adaptation of a structure entrenched with a less than adequate treatment of definitions [2]. According to Arp [2], a step in constructing a domain ontology is to "Determine the most basic universals and Relations, Clearly and Coherently Defining Them." Arp [2] is concerned about a philosopher's ontology that serves primarily to mine and query information that one would typically find on a philosopher's résumé or curriculum vitae. Accordingly, a list of some of the universals and relationships dealt with in this domain are given as follows:

Universals:

person: = def. a human being that is conscious and the full bearer of rights and privileges in a society.
philosopher: = def. a role that a person has whereby that person is considered a practitioner of philosophy.
philosophy: = def. a discipline that studies matters concerning logic, metaphysics, epistemology, ethics, political philosophy, and their associated sub-disciplines [2].

Figure 12 shows these definitions in hierarchical form. Figure 13 shows the FM representation.

**Fig. 12** Hierarchy related to the given definitions

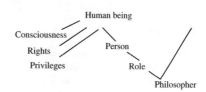

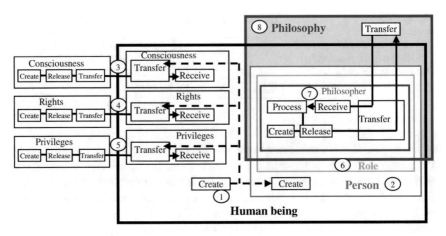

**Fig. 13** FM representation of the given definitions

In the FM representation, there is a human being (circle 1) that is created (appears in the domain). This creation triggers the appearance of a person (2). Consciousness, rights, and privileges flow into (are properties of) that human being (3, 4, and 5). A role appears (6) for the person as a philosopher (7). The philosopher practices philosophy, i.e., he or she transfers, releases, receives, processes, and creates philosophy (8).

Accordingly, it seems from contrasting the textual definitions with the FM representation that FM conveys a richer description involving structure and processes. Additionally, some terms are used with more precision; e.g., *practicing philosophy* can now be defined as transferring, releasing, receiving, processing, and creating philosophy.

## 6  Conclusion

This paper has explored the possiblity of adding a more systematic foundation to philosophical arguments that visualizes structure and incorporates procedures into ontologies. The paper proposes that the notion of flow can serve as a tool to support the hierarchical assembly of terms in the ontology. Such a mixture of structure and flows seems to be a promising approach to underlying any reasoning process instead of the rigid addition of more edges to the hierarchy of ontology.

## References

1. Luenberger, D.: Information Science. Princeton University Press, Princeton, NJ (2006)
2. Arp, R.: Ontology: not just for philosophers anymore. Practical Philos. **10**(1) (2010)

3. Domingue, J., Motta, E.: A knowledge-based news server supporting ontology-driven story enrichment and knowledge retrieval. In: Fensel, D., Studer, R. (eds.) Knowledge Acquisition, Modeling and Management, pp. 104–112. Springer, Berlin (1999)
4. Stalker, I.D., Mehandjiev, N., Carpenter, M.: Devolved ontology for smart applications. In: U., Polovina, S., Hill, R. (eds.) Conceptual Structures: Knowledge Architectures for Smart Applications, Priss, 15th International Conference on Conceptual Structures, ICCS 2007, Sheffield, UK (2007)
5. Gruber, T.: A translation approach to portable ontologies. Knowl. Acquis. **5**, 199–220 (1993)
6. Smith, B.: Ontology. In: Floridi, L. (ed.) Blackwell Guide to the Philosophy of Computing and Information, pp. 155–166. Blackwell, Malden, MA (2003)
7. Arp, R.: Philosophical ontology, domain ontology, formal ontology. In: Williamson, J., Russo, F. (eds.) Key Terms in Logic. Continuum Press, London, pp. 122–123 (2009)
8. Luger, G.: Artificial Intelligence: Structures and Strategies for Complex Problem Solving. Addison-Wesley, New York (2008)
9. Al-Fedaghi, S.: Schematizing proofs based on flow of truth values in logic. IEEE International Conference on Systems, Man, and Cybernetics (IEEE SMC 2013). Manchester, UK (2013)
10. Al-Fedaghi, S.: A method for modeling and facilitating understanding of user requirements in software development. J. Next Gener. Inform. Tech. **4**(3), 30–38 (2013)
11. S. Al-Fedaghi, Anatomy of personal information processing: application to the EU privacy directive, Int. J. Liability Sci. Enquiry **1**(7) (2007)
12. Al-Fedaghi, S.: Awareness of context of privacy. In: 7th International Conference on Knowledge Management (ICKM2010), Pittsburgh, PA (2010)
13. Al-Fedaghi, S.: Pure conceptualization of computer programming instructions. Int. J. Adv. Comput. Tech. **3**(9), 302–313 (2011)
14. Galitsky, B.A., Kovalerchuk, B., Kuznetsov, S.O.: Learning common outcomes of communicative actions represented by labeled graphs. In: Conceptual Structures: Knowledge Architectures for Smart Applications. Lecture Notes in Computer Science, vol. 4604, pp. 387–400
15. McPartland, K.E.: Prediction and Ontology in Aristotle's Organon, Ph.D. dissertation, Jan 2009, Cornell University
16. Guarino, N., Giaretta, P.: Ontologies and knowledge bases: towards a terminological clarification. In: Mars, N. (ed.) Towards Very Large Knowledge Bases: Knowledge Building and Knowledge Sharing. IOS Press, Amsterdam (1995)
17. Ackrill, J.L.: Aristotle's Categories and De Interpretatione. The Clarendon Press, Oxford (1963)
18. Koepsell, D., Arp, R., Fostel, J., Smith, B.: Creating a controlled vocabulary for the ethics of human research: towards a biomedical ethics ontology. J. Empir. Res. Hum. Res. Ethics **4**, 43–58 (2009)

# Mobile Component Integration Agent (MCIA) for Social Business Application

Yvette E. Gelogo, Haeng-Kon Kim and Roger Y. Lee

**Abstract** The Social Business solutions are developed through the integration of different components. Distributed software agents offer great promise in building an increasingly pervasive middleware and component technology. Agents are specialized kinds of components that offer great flexibility than traditional components. This study is focus in developing a software agent that could be used to assemble different type of frameworks which were written and built by different developers of different platforms. This paper proposed a Mobile Component Integration Agent (MCIA) for Social Business Application as a software development methodology to simply integrate the different technology building blocks into one web-based solution. We propose a systematic development process for software agent using component and UML. We first developed the agent components specification and modeled it. Based on this, we developed a mobile application for social business application as a case study. We integrate the developed software framework as a module in Drupal content management System.

**Keywords** Social business · CBD · Web-based solution

## 1 Introduction

A Social Business solution is composed of integrated components that functions differently but have one objective. In this view, this paper proposed a Mobile Component Integration Agent (MCIA) for Social Business Application as a soft-

Y.E. Gelogo · H.-K. Kim
School of Information Technology, Catholic University of Daegu, Daegu, Korea
e-mail: yvette@cu.ac.kr

H.-K. Kim
e-mail: hangkon@cu.ac.kr

R.Y. Lee (✉)
Department of Computer Science, Central Michigan University, Michigan, USA
e-mail: lee1ry@cmich.edu

© Springer International Publishing Switzerland 2016
R. Lee (ed.), *Software Engineering, Artificial Intelligence, Networking and Parallel/Distributed Computing*, Studies in Computational Intelligence 653, DOI 10.1007/978-3-319-33810-1_11

139

ware development methodology to simply integrate the different technology building blocks into one web-based solution.

Social Collaboration is an integrated set of tools that enable real-time knowledge sharing, increased productivity and faster innovation. The most effective approach to enabling a Social Business solutions around helping people discover expertise, develop social networks and capitalize on relationships. A Social Business enables its employees—and customers to more easily find the information and expertise they seek. It helps groups of people bind together into communities of shared interest and coordinate their efforts to deliver better business results faster. It encourages, supports and takes advantage of innovation and idea creation and builds on the intelligence of the crowd.

This study aims to design a framework for social business application which will apply a component based development methodology. To these days, there is a need for fast development of mobile application which is platform independent. There is a need for flexible design for Social Business application systems for mobile. The objectives of this study are to develop a framework with component based development methodology, design a social business application as a case study and develop a mobile application as an output of this study.

The contributions of this paper are the development of framework using component based development methodology for fast software integration regardless of the platform. This study designs a new model of social business application using content management system, Drupal and the output of this study is the integration and development of mobile application.

## 2 Background of the Study

### 2.1 Software Reuse

Software engineering has been more focused on original development but it is now recognized that to achieve better software, more quickly and at lower cost, we need to adopt a design process that is based on systematic software reuse.

*Application system reuse*

- The whole of an application system may be reused either by incorporating it without change into other systems (COTS reuse) or by developing application families.

*Component reuse*

- Components of an application from sub-systems to single objects may be reused.

*Object and function reuse*

- Software components that implement a single well-defined object or function may be reused.

## 2.2 Component Based Development

Component-Based Development claims to offer a radically new approach to the design, construction, implementation and evolution of software applications. Software applications are assembled from components from a variety of sources; the components themselves may be written in several different programming languages and run on several different platforms. CBD architecture is being used nowadays and the research on how to make it more efficient is the focus of this study. A component re-used is one of the most convenient ways for the fast software production. There have been many methods on how to do this and it does involve more technical and detailed view. In this paper we tried to integrate the concept of CBD to develop a mobile enterprise application. We believed that enterprise application uses software components that are being re-used repeatedly; hence, component re-used for mass application developments is necessary [1, 2]. Component Based Development (CBD) is popular methodology to develop a mobile component through component re-used. One of the interesting researches is the enterprise mobile application development with CBD.

## 2.3 Social Business Systems

Social business, as the term has been commonly used since, was defined by Nobel Peace Prize laureate Prof. Muhammad Yunus and is described in his books creating a world without poverty [3]. A social business is a company created with the sole purpose of solving a social problem in a financially self-sustainable way. A good social business combines an unwavering focus on meeting social needs with entrepreneurial energy, market discipline, and great potential for replicating and scaling successful enterprises.

As the rapid growth of social networking and mobility has erased some of the boundaries that separated individuals in the past, people increasingly use their relationships with other people to discover and use information to accomplish innumerable tasks. New opportunities for growth, innovation and productivity exist for organizations that encourage people—employees, customers and partners—to engage and build trusted relationships. Individuals are using social networking tools in their personal lives, and many are also incorporating it into their work lives— regardless of whether it's sanctioned by their employers. Astute organizations will embrace social software and find the most effective ways to utilize it to drive growth, improve client satisfaction and empower employees [4].

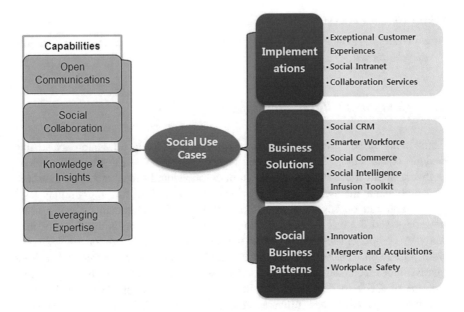

**Fig. 1** Social business use case centric model

Figure 1 show the Social Business use case centric model based on IBM architecture. It has four components, the social collaboration, social analytics, and social content and social user experience. There four components should be meet for successful social business solution. Integrating these components requires different type of technology integration.

There are four main Platforms for Social Business. These are social networking, social analytics, social content, and social user experience [4].

Social networking platforms should be met some considerations

- **Social Networking**

  - People-centric, relationship driven
  - Openness
  - Transparent work and open decision making
  - Connected and discoverable
  - Business driven
  - Adaptable

- **Social Analytics**

  - Infused into social platform
  - Leverage social data to under hidden relationships
  - Make determinations on what people think and might do
  - Integrated solutions

- *Social Content*

  - User Contributed
  - Co-creation
  - Developing content to web, mobile, and social channels
  - Engaging

- *Social User Experience*

  - Role-based, relationship driven social, web, and mobile experiences
  - Dynamic, adaptable, and personal
  - Engaging

# 3   Social Business System Modeling

## 3.1   Component Identification

The component identification stage takes as input the business concept model and the use case model from the requirements workflow. It assumes an application layering that includes a separation of system components and business components. Its goal is to identify an initial set of business interfaces for the business components and an initial set of system interfaces for the system components, and to pull these together into initial component architecture. The business type model is an intermediate artifact from which the initial business interfaces are formed. It is also used later, in the component specification stage, as the raw material for the development of interface information models [5].

Any existing components or other software assets need to be taken into account too, as well as any architecture patterns you plan to use. At this stage it's fairly broad-brush stuff, intended to set out the component and interface landscape for subsequent refinement.

## 3.2   Component Interaction

The component interaction stage examine how each of the system operations will be achieved using the component architecture. It uses interaction models to discover operations on the business interfaces. As more interactions are considered, common operations and patterns of usage emerge that can be factored out and reused. Responsibility choices become clearer and operations are moved from one interface to another. Alternative groupings of interfaces into components can be investigated. This is the time to think through the management of references between component objects so that dependencies are minimized and referential integrity policies are accommodated.

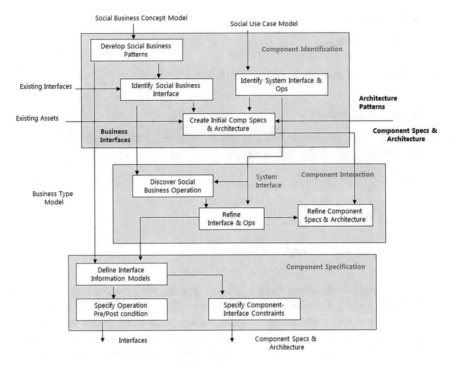

**Fig. 2** Social business specification workflow

The component interaction stage is the stage where the full details of the system structure emerges, with the clear understanding of the dependencies between components, down to the individual operation level.

## 3.3 Component Specification

The final stage of the specification of operations and constraints takes place. For a given interface it means defining the potential states of component objects in an Interface Information Model, and then specifying the pre-and post-conditions for operation and capturing business rules as constraints. This interfaces specification details, witnesses the specification of constraints that are specific to a particular component specification of and independent of each interface. These component specification constraints determine how the type definitions in individual interfaces will correspond to each other in the context of the component.

The architecture should not materially change at this stage. This detailed specification tasks should be undertaken once the architecture is stable and all the operation of the interfaces have been identified. The act of writing the precise rules for each operation may help you discover missing parameters, or missing information, but emphasis is on filling in detail onto a stable framework (Fig. 2).

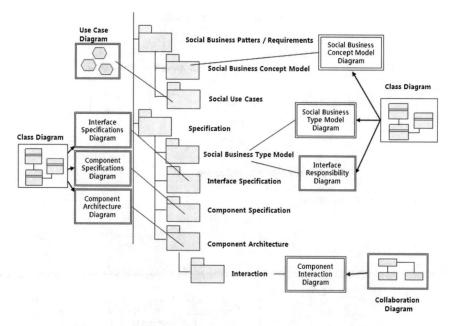

**Fig. 3** Social business component modeling diagram

## 4 Mobile Component Integration Agent (MCIA)

### 4.1 UML Modeling Technique for Social Business System Application

The Social Business Concept Model Diagram is a class diagram depicting the business concept model. An Interface Specification Diagram depicts the interface specification. And so it continues, with the Business Type model Diagram, the Component Specification Diagrams, and the Component Architecture Diagram, each depicting their corresponding artifacts (Fig. 3).

Figure 4 shows the Component Architecture. This is a set of application level software components, their structural relationships, and their behavioral dependencies. A component architecture may apply to a single application or to a wider context, such as a set of applications serving a particular social business process area.

Figure 5 show the Mobile Component Integration Agent (MCIA). The social business enablers are different technology which functions individually. With the component integration process, we come up with one solution. First, we analyze the business process and study the role and relationship of each component.

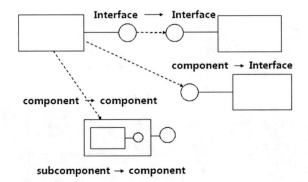

Fig. 4 Component architecture

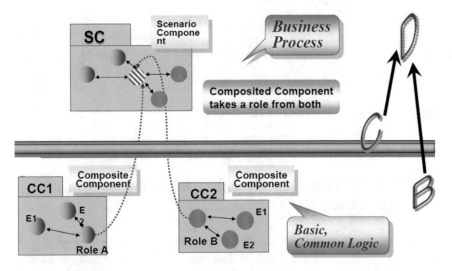

Fig. 5 Mobile component integration agent (MCIA) method

## 5 Case Study

This study proposes a development of social business mobile application using drupal as a case study. Drupal is a content management system (CMS) Drupal allow the user to create a customize module with its own source code to meet the requirements. Drupal makes it very easy to manipulate the content using the hooks. The main node API is implemented by a single hooks that provides a module with the ability to manipulate the node at various stages in its life cycle. The hook gets passed the node object and a parameter indicating what operation is being

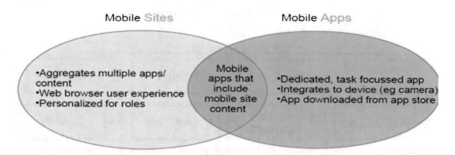

**Fig. 6** Concept of web-based social business solution

performed on it, such us load, update, or view. It is the last of this operation that you want to intercept, because the node is loaded with its HTML content at that point, ready to be rendered, and you can make changes to it if required.

With the use of Content construction Kit (CCK) and views the module is almost part of the CMS as a whole. The CCK module allows site administrator to add extra fields to nodes and augment the standard data model of title, body, author, published date, and so on. These capabilities allow the developer to for flexible design of the web app and the content that needed (Fig. 6).

Figure 7 shows the general overview of the Social Business system for mobile. There are 4 main components design in this model, the profile, communities, meetings and files. We came up with this components based on the social business specifications. The Profile component allows the user to create his/her profile. This also includes messaging, calling, adding friends and notification about the updates. The community components allow the user to create a community, follow other communities, or participate with the discussions and so on. The meeting component is the most complex component. This includes the projects, tasks, reports, memos, request, bulleting and attendance. This is where the meeting is done with other members in the social business. Basically, this is the room for working in the tasks, project and etc. which allow users to collaborate with each other. This component is the biggest component which requires the developer to be more attentive to the integration of different components. Another component that is included is the file. This allows the user to upload files, share it, modify the content and can also be downloaded. This may include spreadsheets and other type of documents.

The tedious task in development of mobile application is the development of hooks and views for social business application. This is where the component development methodology is applied. The way the module is developed in drupal needs the plug-ins and other necessary components for social media application.

Figure 6 shows the concept of web-based solution. This can be access via micro-browser or a mobile application. With a minimal number of taps, users can access their social information. Using the proposed Mobile Component Integration Agent (MCIA) method we came up with the social business system for mobile web-based solution.

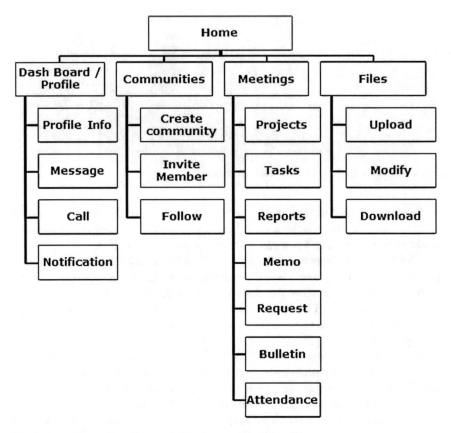

**Fig. 7** General overview of the social business system for mobile

## 5.1 Implementation

The following are needed software and steps to develop the mobile application:

1. Install Web Server (xampp)
2. Install drupal
3. Customize the installed drupal based on web lay-out and contents
4. Work on the needed modules for social business application.

## 6 Conclusion

This study is focus in developing a software agent that could be used to assemble different type of frameworks which were written and built by different developers of different platforms. This paper proposed a Mobile Component Integration Agent

(MCIA) for Social Business Application as a software development methodology to simply integrate the different technology building blocks into one web-based solution. We first developed the agent components specification and modeled it. Based on this, we developed a mobile application for social business application as a case study. We integrate the developed software framework as a module in Drupal content management System as our case study.

**Acknowledgement** This research was Supported by the MSIP (Ministry of Science, ICT and Future Planning), Korea, under the C-ITRC (Convergence Information Technology Research Center) support program (IITP-2015-H8601-15-1007) supervised by the IITP (Institute for Information & communication Technology Promotion).

This research was also supported by the International Research & Development Program of the National Research Foundation of Korea (NRF) funded by the Ministry of Science, ICT & Future Planning (Grant number: K 2013079410).

# References

1. Cepa, V.: Product-line development for mobile device applications with attribute supported containers. PhD Dissertation, Software Technology Group, Technical University of Darmstadt (2005)
2. Kim, H.-K.: Mobile agent development with CBD on ABCD architectures. In: International MultiConference of Engineers and Computer Scientists (2013)
3. Yunus, M.: Creating a world without poverty: social business and the future of capitalism. Public Aff. 320 (2009). ISBN: 978-1-58648-667-9
4. Social business architecture. IBM Corporation (2014). Accessed 10 Mar 2015
5. Cheesman, J., Daniels, J.: UML components: A Simple Process for Specifying Component-Based Software. The Addison-Wesley Object Technology Series (2001)
6. Cheesman, J., Daniels, J.: UML components: A Simple Process for Specifying Component-Based Software. The Addison-Wesley Object Technology Series (2000). ISBN: 0-201-70851-5
7. The Social Business, Advent of a new age, White Paper, EPW14008-USEN-00. https://www.ibm.com/smarterplanet/global/files/us__en_us__socialbusiness__epw14008usen.pdf. Accessed 10 Feb 2015
8. Agner, L.T.W., Soares, I.W., Stadzisz, P.C., Simao, J.M.: Model refinement in the model driven development context. J. Comput. Sci. **8**(8), 1205–1211 (2012). ISSN: 1549-3636
9. Kim, H.-K.: Design of web-based social business solution architecture based on CBD. Int. J. Softw. Eng. Appl. **9**(2), 271–278 (2015)

# Covariance Estimation for Vertically Partitioned Data in a Distributed Environment

**Aruna Govada and S.K. Sahay**

**Abstract** The major sources of abundant data is constantly expanding with the available data collection methodologies in various applications—medical, insurance, scientific, bio-informatics and business. These data sets may be distributed geographically, rich in size and as well as dimensions also. To analyze these data sets to find out the hidden patterns, it is required to download the data to a centralized site which is a challenging task in terms of the limited bandwidth available and computationally also expensive. The covariance matrix is one of the method to estimate the relation between any two dimensions. In this paper we propose a communication efficient algorithm to estimate the covariance matrix in a distributed manner. The global covariance matrix is computed by merging the local covariance matrices using a distributed approach. The results show that it is exactly same as centralized method with good speed-up in terms of computation. The reason for speed-up is because of the parallel construction of local covariances and distributing the cross covariances among the nodes so that the load is balanced. The results are analyzed by considering Mfeat data set on the various partitions which addresses the scalability also.

**Keywords** Parallel/distributed computing · Covariance matrix · Vertical partition

## 1 Introduction

Ongoing projects and future projects in various disciplines like earth sciences, astronomy, climate variability, cancer research (e.g. CORAL, SWOT, WISE, LSST, SKA, JASD, AACR) [1–7] are destined to produce the enormous catalogs which will be geographically distributed. As the amount of data available at various

A. Govada (✉) · S.K. Sahay
BITS-Pilani, K.K. Birla Goa Campus, Zuarinagar 403726, Goa, India
e-mail: garuna@goa.bits-pilani.ac.in

S.K. Sahay
e-mail: ssahay@goa.bits-pilani.ac.in

© Springer International Publishing Switzerland 2016
R. Lee (ed.), *Software Engineering, Artificial Intelligence, Networking and Parallel/Distributed Computing*, Studies in Computational Intelligence 653, DOI 10.1007/978-3-319-33810-1_12

151

geographically distributed sources is increasing rapidly, traditional centralized techniques for performing data analytics are proving to be insufficient for handling this data avalanche [8]. Downloading and processing all the data at a single location results in increased communication as well as infrastructural costs [9].

Bringing these massive data sets which are distributed geographically to a centralized site is almost impossible due to the limited bandwidth when compared with the size of the data. And also solving a problem with large number of dimensions at a central site is not practical as it is computationally expensive. Analyzing these massive data can not be achieved unless the algorithms are capable of handling the decentralized data [8].

These data sets might be distributed in two different ways either horizontally or vertically [10]. In Horizontal partition the number of attributes/dimensions are constant at all n different locations but the number of instances may vary. Whereas in vertical partition the number of instances are constant at all n different locations but number of dimensions may vary. In this paper the data is partitioned in vertical manner.

The analysis of these vertically partitioned geographically distributed data sets assume that the data should fit into main memory which is a challenge task in terms of scalability. Estimation of covariance matrix analyses how the data is related among the dimensions. The task of estimating the covariance matrix of the data sets demand the data to be available at one centralized site [15].

In this paper covariance matrix is estimated for vertically partitioned data in a decentralized manner without brining the data to a centralized site. The proposed distributed approach is compared with the centralized method by bringing the distributed data to one central site. The estimation of covariance matrix is achieved both in centralized and distributed approach. The experimental analysis shows how our distributed approach is better than the normal approach in terms of speed-up with exactly same solution. Results are analyzed by considering various partitions of Mfeat data set [18].

The rest of the organization of the paper is as follows. Section 2 introduces the related work. In Sect. 3 preliminaries and notations are briefly described. In Sect. 4 we present our distributed approach for distributed covariance matrix (DCM) and also discusses the speed-up of our approach when compared with centralized version. In Sect. 5 we present the experimental analysis of our algorithm. At the end in Sect. 6 the conclusions of the paper are mentioned.

## 2 Related Work

Estimation of covariance based on divide and conquer approach is discussed by Nik et al. in which the computational cost is reduced [11]. A regularization and blocking estimator of high dimensional covariance is discussed by et al. using Barndorff Nielson Hansen estimator [12]. Modified Cholesky decomposition and other decomposition methods are discussed for the estimation of covariance by Hao for high dimensional data with limited sample size [13]. Guo et al. proposed a divide conquer approach based on feature space decomposition for classification [14]. The significance of distributed estimation of parameters over centralized method is discussed and belief propagation algorithm is investigated by Jain [15]. l1-regularized Gaussian maximum likelihood estimator (MLE) is discussed by Cho et al. in recovering a sparse inverse covariance matrix for high-dimensional data which statistically guarantees using a single machine [16]. Aruna et al. discussed the distributed approach for multi classification using SVM without bringing data to a centralized site [17].

## 3 Preliminaries

### 3.1 Covariance

The statistical analysis of the data sets usually investigates the dimensions, to see if there is any relationship between them. covariance is the measurement, to find out how much the dimensions vary from the mean with respect to each other.

The covariance of two dimensions X, Y can be compute as

$$cov(X, Y) = \frac{\sum_{i=1}^{i=n}(X_i - \mu_x)(Y_i - \mu_y)}{n - 1}$$

where $\mu_x$ and $\mu_y$ are the mean of the dimensions X and Y respectively.

### 3.2 Covariance Matrix

Covariance is always computed between the two dimensions. If the data contains more than two dimensions, there is a requirement to calculate more than one covariance measurement.

The standard way to get the covariance values between the different dimensions of the data set is to compute them all and put them in a matrix. The covariance matrix for a set of data with k dimensions is:

$$C_{k \times k} = (c_{i,j}, c_{i,j} = cov(Dim_i, Dim_j))$$

where $C_{k \times k}$ is a matrix with $k$ rows and $k$ columns, and $Dim_i$ is the $i$th dimension. If we have an $k$-dimensional data set, then the matrix is a square matrix of $k$ dimensions and each value in the matrix is the computed covariance between two distinct dimensions.

Consider for an imaginary $k$ dimensional data set, using the dimensions $l_1, l_2, l_3 \ldots l_k$, Then, the covariance matrix has $k$ rows and $k$ columns, and the values are:

The *covariance Matrix* $C_{k \times k}$ is an $k \times k$ matrix which can be written as follows.

$$\begin{pmatrix} l_1 l_1 & l_1 l_2 & l_1 l_3 & \cdots & \cdots & l_1 l_k \\ l_2 l_1 & l_2 l_2 & l_2 l_3 & \cdots & \cdots & l_2 l_k \\ \cdots & \cdots & \cdots & \cdots & \cdots & \cdots \\ \cdots & \cdots & \cdots & \cdots & \cdots & \cdots \\ l_k l_1 & l_k l_2 & l_k l_3 & \cdots & \cdots & l_k l_k \end{pmatrix}$$

Along the main diagonal, the covariance value is between one of the dimensions and itself. These are nothing but the variances for that dimension.

The other point is that since $cov(l_1, l_2) = cov(l_2, l_1)$ the matrix is symmetrical about the main diagonal.

## 4  The Proposed Approach

### 4.1  Distributed Covariance Matrix (DCM)

The data be distributed among t sites with equal number of instances but varied in number of dimensions i.e. vertically partitioned data.

1. Let the data is distributed among t sites and the sites are labeled as $S_0, S_1, S_{t-1}$.

$$[\mathbf{X}]_{l \times m} = (X_0, X_1, X_2, \ldots X_{t-1})$$

where data $X_j$ is a $l \times m_j$ matrix residing at the site $S_j$ and $m = \sum_{j=0}^{t-1} m_j$.

2. Calculate the local covariances $C_{00}, C_{11} \ldots C_{t-1t-1}$ at all t sites parallely.
3. If the number of sites are only 2, Either send the corresponding data from $S_0$ to $S_1$ or from $S_1$ to $S_0$ and calculate the cross covariances.
4. If the number of sites are more than 2, Calculate the cross covariances $C_{jk}$ by sending the corresponding data $X_j$ of $S_j$ to the site $S_k$ as follows.

   • If the number of sites are even, $t = 2r$

     – for $k = 0$ to $r - 1$

       $j$ = immediate $r - 1$ predecessor sites

     – for $k = r$ to $t - 1$

       $j$ = immediate $r$ predecessor sites

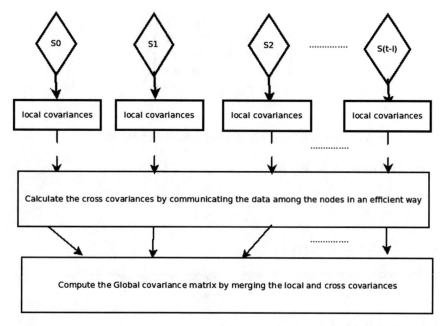

**Fig. 1** The architecture

- If the number of sites are odd, $t = 2r + 1$

  - for $k = 0$ to $t - 1$

    $j$ = immediate $r$ predecessor sites

5. Merge the local and cross covariances to get the global covariance matrix.
6. Estimate the eigen components of the global covariance matrix.

The architecture of the proposed approach is shown in Fig. 1, where the global covariance matrix is computed by merging the local and cross covariances.

### 4.2  Global Covariance Matrix

Let us consider 3 nodes $n_0$, $n_1$, $n_2$. The node $n_0$ consists of two columns labeled by x, y. The node $n_1$ also consists of two columns labeled by z, w. The node $n_2$ consists of single column labeled by v. The covariance matrix by centralized approach would be (considering only upper triangular matrix as covariance is symmetric):

$$\begin{pmatrix} xx & xy & xz & xw & xv \\ - & yy & yz & yw & yv \\ - & - & zz & zw & zv \\ - & - & - & ww & wv \\ - & - & - & - & vv \end{pmatrix}$$

### 4.2.1 Computation of Global Covariance Matrix by DCM

Local Covariance of $n_0$, say $lc_0$

$$\begin{pmatrix} xx & xy \\ - & yy \end{pmatrix}$$

Local Covariance of $n_1$, say $lc_1$

$$\begin{pmatrix} zz & zw \\ - & ww \end{pmatrix}$$

Local Covariance of $n_2$, say $lc_0$

$$( vv )$$

Cross Covariance of $n_0$ and $n_1$, say $cc_{01}$

$$\begin{pmatrix} xz & xw \\ yz & yw \end{pmatrix}$$

Cross Covariance of $n_1$ and $n_2$, say $cc_{12}$

$$\begin{pmatrix} zv \\ wv \end{pmatrix}$$

Cross Covariance of $n_0$ and $n_2$, say $cc_{02}$

$$\begin{pmatrix} xv \\ yv \end{pmatrix}$$

Global Covariance matrix by merging the local and cross covariances as given below would be equivalent to the matrix calculated by centralized approach.

$$\begin{pmatrix} lc_0 & cc_{01} & cc_{02} \\ - & lc_1 & cc_{12} \\ - & - & lc_2 \end{pmatrix}$$

### 4.3 The Efficient Communication Among the Nodes

The data is communicated among the sites in such a manner so that the resources are used in an efficient way. The computational load is also balanced among the sites to have the good speed-up. When the number of sites are even i.e. $2r$, the first $r$ sites will receive the data from their immediate $(r-1)$ predecessors. Then the remaining $r$ sites will receive the data from their immediate $r$ predecessors. Sharing of this data by communicating among the sites is illustrated in Fig. 2, when the number of sites say t = 4. Here the value of r = 2. So the first 2 sites $S_0$ and $S_1$ will receive the data from its immediate $(r-1)$ predecessors i.e. $S_0$ will receive data from $S_3$ and $S_1$ will receive data from $S_0$. The next 2 sites $S_2$ and $S_3$ will receive the data from its immediate $r$ predecessors i.e. $S_2$ will receive the data from $S_1$, $S_0$ and $S_1$ will receive the data from $S_0$, $S_4$.

When the number of sites are odd, all the $(2r+1)$ sites will receive the data from their immediate $r$ predecessors. Sharing of this data by communicating among the sites is illustrated in Fig. 3, when the number of sites say t = 5. Here the value of r = 2. So all the 5 sites from $S_0$ to $S_4$ will receive the data from its immediate $r$ predecessors i.e. $S_0$ will receive the data from $S_4$ and $S_3$, $S_1$ will receive the data from $S_0$ and $S_4$, $S_2$ will receive the data from $S_1$ and $S_0$, $S_3$ will receive data from $S_2$

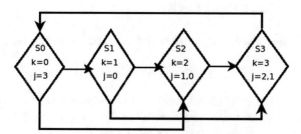

**Fig. 2** The number of nodes are say 4 (even): sending the data of $j$th site to $k$th site

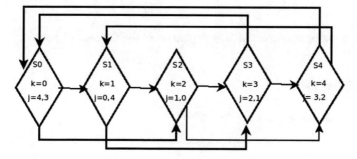

**Fig. 3** The number of nodes are say 5 (odd): sending the data of $j$th site to $k$th site

and $S_1$. Therefore the number of transfers of the sites of the data will be at most r in all the cases. It is not required that all the sites should have the data of remaining sites.

## 4.4 Speed-Up of DCM

### 4.4.1 Computational Time of Centralized Approach

In centralized version let the data is available in a single matrix

$$[\mathbf{X}]_{l \times m} = (X_0, X_1, X_2, \ldots X_{t-1})$$

where data $X_j$ is a $l \times m_j$ matrix residing at the site $S_j$ and $m = \sum_{j=0}^{t-1} m_j$ Let the computational time of centralized approach is denoted as $T_c$

$$T_c = \frac{m(m-1)}{2}$$

### 4.4.2 Computational Time of DCM

As the data is distributed among t sites and the sites are labeled as $S_0, S_1, S_{t-1}$.

$$[\mathbf{X}]_{l \times m} = (X_0, X_1, X_2, \ldots X_{t-1})$$

Let computational time of global/distributed covariance matrix is denoted as $T_d$, computational time of (local covariances) as $T_l$, computational time of (Cross covariances) as $T_{cr}$ and the communication cost as $T_{cm}$

$$T_d = T_l + T_{cr} + T_{cm}$$
$$= Max\left(\frac{m_j(m_j - 1)}{2}\right) + Max\left(\sum_{k=0}^{k=t-1} \sum_{i} ((m_k \times m_i) + m_i)\right)$$

where i is the predecessors of k as explained in the 4th step of Sect. 4.1.

### 4.4.3 Speed-Up

Let us denote the Speed-Up by S

$$S = \frac{T_c}{T_d}$$

$$= \frac{\frac{m(m-1)}{2}}{Max\left(\frac{m_j(m_j-1)}{2}\right) + Max\left(\sum_{k=0}^{k=t-1} \sum_i \left((m_k \times m_i) + m_i\right)\right)}$$

Consider the t sites with each of $\Gamma$ columns of data.

$$T_c = \frac{(t\Gamma)(t\Gamma - 1)}{2}$$

$$T_l = \frac{(\Gamma)(\Gamma - 1)}{2}$$

$$T_{cr} = \Gamma r, \quad T_{cm} = \Gamma r$$

$$T_d = \frac{\frac{(t\Gamma)(t\Gamma-1)}{2}}{\frac{(\Gamma)(\Gamma-1)}{2} + \Gamma r + \Gamma r}$$

$$= \frac{t(t\Gamma - 1)}{\Gamma - 1 + 4r}$$

**Case1:** $t = 2r$ (even)

$$= \frac{(2r)(2r\Gamma - 1)}{(\Gamma - 1) + 4r}$$

$$= \frac{4r^2\Gamma - 2r}{4r + \Gamma - 1}$$

**Case2:** $t = 2r + 1$ (odd)

$$= \frac{(2r+1)((2r+1)\Gamma - 1)}{(\Gamma - 1) + 4r}$$

$$= \frac{4r^2\Gamma + (1+4r)\Gamma - 2r - 1}{4r + \Gamma - 1}$$

In both the cases speed-up will be at least $r$ times.

## 5   Experimental Analysis

We implemented the algorithm with the data set Mfeat, taken from UCI machine learning repository https://archive.ics.uci.edu/ml/datasets.html. Mfeat data consists of 2000 rows and are distributed in six data files as follows [18]:

1. Mfeat-fac: 216 profile correlations;
2. Mfeat-fou: 76 Fourier coefficients of the character;
3. Mfeat-kar: 64 Karhunen Love coefficients;
4. Mfeat-mor: 6 morphological features;
5. Mfeat-pix: 240 pixel averages in 2 × 3 windows;
6. Mfeat-zer: 47 Zernike moments.

The algorithm is implemented using Java Agent DEvelopment framework (JADE) [19]. Each site data is downloaded to a node which are connected over the network. So the number of computational nodes is equal to the number of sites. The communication is established among them using JADE to transfer the data.

In our analysis the vertical partitions are considered from 2 to 6 which is shown in Table 1. The computational time of local and cross covariances are shown in

**Table 1**  The various partitions considered for distributed computation

| Dataset | Rows | Cols | No. of partitions | Cols considered at each node/site |
|---------|------|------|-------------------|-----------------------------------|
| Mfeat   | 2000 | 648  | 2 | Fact-Fou-Kar, Mor-Pix-Zer |
|         |      |      | 3 | Fact, Fou-Kar, Mor-Pix-Zer |
|         |      |      | 4 | Fact, Fou-Kar, Mor-Pix, Zer |
|         |      |      | 5 | Fact, Fou, Kar, Mor-Pix, zer |
|         |      |      | 6 | Fact, Fou, Kar, Mor, Pix, zer |

**Table 2**  Distributed computational time when number of partitions = 6

| Dataset | Local covariances | Cross covariances | | |
|---------|-------------------|-------------------|---|---|
| $S_0$: Fact | $S_0S_0$: 3439 | $S_0S_5$: 1500 | $S_0S_4$: 3165 | – |
| $S_1$: Fou | $S_1S_1$: 708 | $S_1S_5$: 796 | $S_1S_0$: 1877 | – |
| $S_2$: Kar | $S_2S_2$: 684 | $S_2S_1$: 896 | $S_2S_0$: 1301 | – |
| $S_3$: Mor | $S_3S_3$: 250 | $S_3S_2$: 526 | $S_3S_1$: 488 | $S_3S_0$: 804 |
| $S_4$: Pix | $S_4S_4$: 3822 | $S_4S_3$: 647 | $S_4S_2$: 1963 | $S_4S_1$: 1965 |
| $S_5$: Zer | $S_5S_5$: 528 | $S_5S_4$: 1548 | $S_5S_3$: 436 | $S_5S_2$: 749 |

**Table 3**  Distributed computational time when number of partitions = 5

| Dataset | Local covariances | Cross covariances | |
|---------|-------------------|-------------------|---|
| $S_0$: Fact | $S_0S_0$: 3439 | $S_0S_4$: 1500 | $S_0S_3$: 3142 |
| $S_1$: Fou | $S_1S_1$: 708 | $S_1S_4$: 796 | $S_1S_0$: 1877 |
| $S_2$: Kar | $S_2S_2$: 684 | $S_2S_1$: 896 | $S_2S_0$: 1301 |
| $S_3$: Mor-Pix | $S_3S_3$: 4186 | $S_3S_2$: 1445 | $S_3S_1$: 2081 |
| $S_4$: Zer | $S_4S_4$: 528 | $S_4S_3$: 1543 | $S_4S_2$: 749 |

**Table 4** Distributed computational time when number of partitions = 4

| Dataset | Local covariances | Cross covariances | |
|---|---|---|---|
| $S_0$: Fact | $S_0S_0$: 3439 | $S_0S_3$: 1500 | – |
| $S_1$: Fou-Kar | $S_1S_1$: 1415 | $S_1S_0$: 2354 | – |
| $S_2$: Mor-Pix | $S_2S_2$: 4186 | $S_2S_1$: 3400 | $S_2S_0$: 3142 |
| $S_3$: Zer | $S_3S_3$: 528 | $S_3S_2$: 1543 | $S_3S_1$: 1013 |

**Table 5** Distributed computational time when number of partitions = 3

| Dataset | Local covariances | Cross covariances |
|---|---|---|
| $S_0$: Fact | $S_0S_0$: 3439 | $S_0S_2$: 3542 |
| $S_1$: Fou-Kar | $S_1S_1$: 1415 | $S_1S_0$: 2354 |
| $S_2$: Mor-Pix-Zer | $S_2S_2$: 3704 | $S_2S_1$: 2108 |

**Table 6** Distributed computational time when number of partitions = 2

| Dataset | Local covariances | Cross covariances |
|---|---|---|
| $S_0$: Fact-Fou-Kar | $S_0S_0$: 3570 | $S_0S_1$: 2561 |
| $S_1$: Mor-Pix-Zer | $S_1S1$: 3704 | – |

**Table 7** The communication cost of sending predecessors (in milliseconds)

| No. of partitions | Site | Predecessors cost |
|---|---|---|
| 6 | $S_0$: Fact | $S_5$: 430 $S_4$: 2165 |
| | $S_1$: Fou | $S_0$: 1950 $S_1$: 430 |
| | $S_2$: Kar | $S_1$: 685 $S_0$: 1950 |
| | $S_3$: Mor | $S_2$: 570 $S_1$: 685 $S_0$: 1950 |
| | $S_4$: Pix | $S_3$: 45 $S_2$: 570 $S_1$: 685 |
| | $S_5$: Zer | $S_3$: 45 $S_2$: 570 $S_1$: 685 |
| 5 | $S_0$: Fact | $S_4$: 430 $S_3$: 2240 |
| | $S_1$: Fou | $S_0$: 1950 $S_4$: 430 |
| | $S_2$: Kar | $S_1$: 685 $S_0$: 1950 |
| | $S_3$: Mor-Pix | $S_2$: 570 $S_1$: 685 |
| | $S_4$: Zer | $S_3$: 2240 $S_2$: 570 |
| 4 | $S_0$: Fact | $S_3$: 430 |
| | $S_1$: Fou-Kar | $S_0$: 1950 |
| | $S_2$: Mor-Pix | $S_1$: 1280 $S_4$: 1950 |
| | $S_3$: Zer | $S_2$: 2240 $S_4$: 1280 |
| 3 | $S_0$: Fact | $S_2$: 2660 |
| | $S_1$: Fou-Kar | $S_0$: 1950 |
| | $S_2$: Mor-Pix-Zer | $S_1$: 1280 |
| 2 | $S_0$: Fact-Fou-Kar | $S_1$: 2660 |
| | $S_1$: Mor-Pix-Zer | – |

Tables 2, 3, 4, 5 and 6 for the partitions 6, 5, 4, 3, 2 respectively. The cross covariances are chosen as explained in Sect. 4.1, step 4. In Table 7 the communication cost for a given site for sending its predecessors data is shown. In Table 8,

**Table 8** Comparison of computational time (in milliseconds) of centralized and distributed versions

| Dataset | No. of partitions | Centralized | Distributed |
|---------|-------------------|-------------|-------------|
| 3* Mfeat | 2 | 8855 | 8791 |
| | 3 | 9937 | 9641 |
| | 4 | 15,311 | 13,958 |
| | 5 | 15,582 | 11,498 |
| | 6 | 18,486 | 9347 |

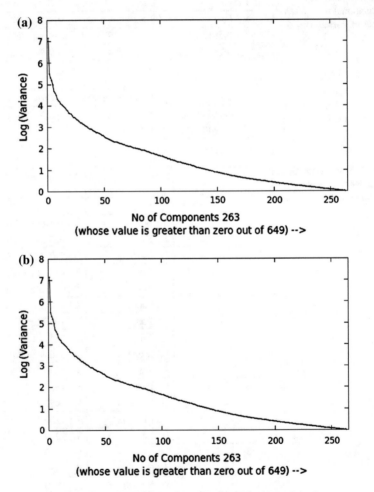

**Fig. 4** Covariance estimations of Mfeat data set. **a** Centralized. **b** Distributed

the computational time of centralized and distributed approaches are compared. The computational time of distributed approach is calculated from Tables 2, 3, 4, 5 and 6 and from Table 7 for various partitions as explained in Sect. 4.2.1. In our analysis DCM is compared with centralized approach, the result is exactly same as shown in

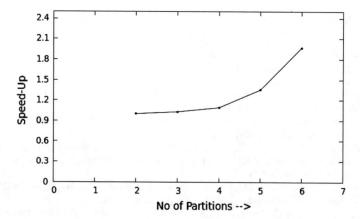

**Fig. 5** Speed-up of DCM

Fig. 4. Because we are not losing any data but getting the distributed covariance matrix by merging the local and cross covariances. The speed-up is shown in Fig. 5. It is observed that the speed-up is increasing with the number of partitions hence scalable. This is because of increase in parallel computations along with the number of partitions. There is an elevation in speed up when the number of partitions are ≥5 which promises that it works well even with number of partitions are increasing.

# 6  Conclusions

We propose an algorithm DCM which estimates the global covariance matrix by merging the local and cross covariances that are distributed at different nodes/sites. Experimental results show that the result of DCM is exactly same as the centralized approach with good speed-up. The final output of DCM is same as centralized approach because we are not losing any data. The computational time of DCM is decreasing along the increased number of partitions. DCM is also capable of handling large data sets based on parallel calculations of vertical partitions, hence scalable. The speed-up can be further increased by making number of columns equal at every node/site and also computing the cross covariances parallely within the node/site.

**Acknowledgments** We are thankful for the support provided by the Department of CSIS, BITS-Pilani, K.K. Birla Goa Campus to carry out the experimental analysis and also to Sreejith.V, BITS-Pilani, K.K. Birla Goa Campus for useful discussions.

# References

1. http://science.nasa.gov/missions/coral/
2. https://swot.jpl.nasa.gov/mission/
3. http://www.jpl.nasa.gov/wise
4. Large Synoptic Survey Telescope: http://www.lsst.org
5. https://www.skatelescope.org/project
6. http://science.nasa.gov/about-us/smd-programs/joint-agency-satellite-division/
7. http://www.aacr.org/AboutUs/Pages/default.aspx
8. Bhaduri, K., Das, K., Borne, K., et al.: Scalable, asynchronous, distributed eigen-monitoring of astronomy data streams. In: Proceedings of the SIAM International Conference on Data Mining, pp. 247–258 (2009)
9. Weske, M., Shacid, M., Godart, C. (eds.): Data in astronomy: from the pipeline to the virtual observatory. WISE Workshops, LNCS 4832, pp. 52–62 (2007)
10. Dutta, H., Giannella, C., Borne, K., et al.: Distributed Top-K outlier detection from astronomy catalogs using the DEMAC system. In: Proceedings of SDM07, pp. 473–478 (2007)
11. Hsieh, C.J., Dhillon, I.S., Ravikumar, P., Banerjee, A.: A divide-and-conquer procedure for sparse inverse covariance estimation. In: Advances in Neural Information Processing Systems, pp. 2330–2338 (2012)
12. Hautsch, N., Kyj, L.M., Oomen, R.C.A.: A blocking and regularization approach to high-dimensional realized covariance estimation. J. Appl. Econ. **27**(4), 625–645 (2012)
13. Hao, Z.: Large Dimensional Covariance Matrix Estimation with Decomposition-based Regularization, p. 129. https://books.google.co.in/books?id=SsL2jgEACAAJ (2014)
14. Guo, Q., Chen, B.-W., Jiang, F., Ji, X., Kung, S.-Y.: Efficient divide-and-conquer classification based on feature-space decomposition. IEEE Syst. J.
15. Du, J., Ng, T.S., Wu Y.: Distributed estimation in large-scale networks: theories and applications. http://hdl.handle.net/10722/197090 (2013)
16. Hsieh, C.-J., Sustik, M.A., Dhillon, I.S., Ravikumar, P., Poldrack, R.A.: Russell sparse inverse covariance estimation for a million variables. In: Advances in Neural Information Processing Systems (2013)
17. Govada, A., Gauri, B., Sahay, S.K.: Distributed multi-class SVM for large data sets. In: Proceedings of the Third International Symposium on Women in Computing and Informatics, Cochi, India, pp. 54–58. ACM (2015)
18. Mfeat Data set on UCI Machine Learning Repository: https://archive.ics.uci.edu/ml/datasets/Multiple+Features
19. Java Agent DEvelopment framework: http://jade.tilab.com

# Detection of Dengue Epidemic in Dhaka, Bangladesh by a Neuro Fuzzy Approach

Md. Arifuzzaman, Md. Faqrul Islam Shaon, Md. Jahidul Islam
and Rashedur M. Rahman

**Abstract** The aim of this research is to develop an efficient system which will identify the probability of dengue occurrence in Dhaka, Bangladesh based on a neural network system and fuzzy inference algorithm. When using fuzzy inference technique, we separated the dengue cases into four quadrants of 3 months in a year. Based on dengue infection rate each quadrant is classified from high to low. An adaptive neuro based fuzzy inference system provides the insight for implementing fuzzy rules. Then we analyze the performances using fuzzy logic and ANFIS to point out the percentage of infected rate.

## 1 Introduction

Dengue Fever (DF) is a major health related problem in Dhaka, Bangladesh. It is more of a dangerous and debilitating disease which is a growing threat over the countryside. It was found that due to erratic water supply and transmission of dengue virus the infection is moving fast. However, drawing a vast area of research based on infection rate the probability could be predetermined by using fuzzy approach. This paper presents a significant model in such a way that just by giving

Md. Arifuzzaman · Md. Faqrul Islam Shaon · Md. Jahidul Islam · R.M. Rahman (✉)
Department of Electrical and Computer Engineering, North South University,
Bashundhara, Dhaka, Bangladesh
e-mail: rashedur.rahman@northsouth.edu

Md. Arifuzzaman
e-mail: zamanarif93@gmail.com

Md. Faqrul Islam Shaon
e-mail: md_shaon@outlook.com

Md. Jahidul Islam
e-mail: zahidtanmay009@gmail.com

© Springer International Publishing Switzerland 2016
R. Lee (ed.), *Software Engineering, Artificial Intelligence, Networking and Parallel/Distributed Computing*, Studies in Computational Intelligence 653, DOI 10.1007/978-3-319-33810-1_13

the time frame of the infected people it will identify dengue epidemic. Instead of crisp set it will use fuzzy set for flexibility. Then adaptive Neuro Fuzzy Inference System (ANFIS) was used to pin point the infected epidemic rate.

The rest of the paper is organized as follows. Section 2 will briefly discuss a number of related works in this area. Section 3 derives the data acquisition part. Section 4 describes methodology. Section 5 analyzes the performance result. Section 6 gives conclusion and future work regarding this project.

## 2   Related Work

Fuzzy logic has been used in many areas that include disease detection and epidemic predictability such as diarrhoea, typhoid fever etc. One of the earliest implementation of fuzzy based algorithm is detection of Parkinson's disease. It uses fuzzy k-nearest method to identify Parkinson's disease. The causes of this disease are measured on vocal impairment symptoms [1]. There is also disease detection like orchid disease detection implemented based on fuzzy algorithm. Orchid disease detection uses image processing to identify the symptoms and fuzzy logic to identify infection rate [2]. An implementation of fuzzy based algorithm on cholera found that diseases like cholera occurred in Bangladesh almost very often but the outbreaks are not the same in every year. It was also found that the virus could be spread due to water temperature and environmental issues [3]. A research done in Vietnam showed that dengue magnitude can be forecasted a week ahead in advance by taking statistical data and biological factors as input in a locality and transmission model. They also used an artificial neural network to predict the mobility of dengue in provinces of Vietnam [4].

## 3   Data Acquisition

We collect the dataset manually from Institute of Epidemiology, Disease Control and Research (IEDCR) and Centre of Disease Control (CDC) annual report Bangladesh (version 28/05/2013) which is available online [5]. The dataset is designed with 9 consecutive years of infected people of Dhaka, Bangladesh starting from 2007 to 2015 [6]. Figure 1 shows the people infected during those 9 consecutive years including each month. Figure 2 shows the total population during those years in Dhaka, Bangladesh. Figure 3 shows the infection rate of Dengue in Dhaka.

| | A | B | C | D | E | F | G | H | I | J |
|---|---|---|---|---|---|---|---|---|---|---|
| 1 | | 2007 | 2008 | 2009 | 2010 | 2011 | 2012 | 2013 | 2014 | 2015 |
| 2 | January | 0 | 0 | 0 | 0 | 0 | 0 | 0 | 0 | 0 |
| 3 | February | 0 | 0 | 0 | 0 | 0 | 0 | 0 | 0 | 0 |
| 4 | March | 0 | 0 | 0 | 0 | 0 | 0 | 2 | 0 | 2 |
| 5 | April | 0 | 0 | 0 | 0 | 0 | 0 | 6 | 0 | 6 |
| 6 | May | 0 | 0 | 0 | 0 | 10 | 3 | 11 | 2 | 10 |
| 7 | June | 0 | 4 | 0 | 15 | 30 | 6 | 25 | 11 | 28 |
| 8 | July | 81 | 169 | 2 | 171 | 207 | 156 | 171 | 28 | 171 |
| 9 | August | 195 | 577 | 140 | 195 | 737 | 135 | 365 | 155 | 765 |
| 10 | Septembe | 170 | 235 | 175 | 28 | 175 | 225 | 381 | 57 | 965 |
| 11 | October | 20 | 168 | 147 | 0 | 147 | 128 | 503 | 21 | 869 |
| 12 | Novembe | 0 | 0 | 10 | 0 | 35 | 21 | 225 | 6 | 266 |
| 13 | Decembei | 0 | 0 | 0 | 0 | 21 | 0 | 55 | 0 | 21 |

**Fig. 1** Infected people

| Year | Population |
|---|---|
| 2007 | 9,997,843 |
| 2008 | 10,461,728 |
| 2009 | 10,921,290 |
| 2010 | 11,511,392 |
| 2011 | 12,043,977 |
| 2012 | 12,651,104 |
| 2013 | 13,173,159 |
| 2014 | 13,862,584 |
| 2015 | 14,543,124 |

**Fig. 2** Number of population

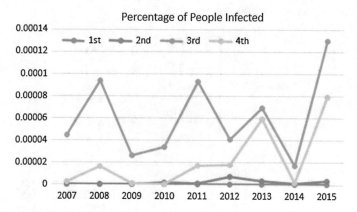

**Fig. 3** Infection rate of dengue in Dhaka

## 4  Methodology

In the first stage, the months are divided into four quadrants. Since we know dengue only occurs in certain months this division allowed us to discard the possibility of epidemic in certain months. Using quadrants of months instead of seasons we can detect the membership functions distinctly [7]. The distinction underlines the endemic concept. Endemic refers to characteristic that is constantly present to a greater or lesser degree in people of a certain class or in people living in a particular geographic location. This level we are seeking is the observed level which is not necessary to be the desired level i.e. to be zero. Dengue might continue to spread at this level considerably due to absence of intervention. Thus, the base line level can be considered as the expected level in these criteria. Afterwards we construct our Fuzzy Inference System by considering the variables. We then run our input variables over ANFIS to construct our Adaptive Neuro Fuzzy Inference System to get better accuracy. Finally, we'll defuzzify all and get the percentage.

### 4.1  Classification of 4 Quadrants

We consider 3 months for each quadrant to make our membership function. Bangladesh is a country of six seasons. Although other common diseases occur all the year round, dengue occurs in some particular months only therefore the division of months in quadrants makes it easier. So we classify all the months into four quadrants with each consist of 3 months (consecutive). Then we use Gaussian Membership function to separate the points of the quadrants. In order to maintain a wider insight, we have chosen the standard deviation to be around 2 months long i.e., 51 for our membership functions. Equation 1 depicts the membership functions (Table 1 and Fig. 4).

Quadrant Membership Equation

$$\mu_{quadrant}(\mathbf{X}) = \begin{cases} e^{-\frac{(x+3.97e-15)^2}{2(51.67)^2}}, & 0 \le x \le 61 \\ e^{-\frac{(x-121.3)^2}{2(51.67)^2}}, & 61 \le x \le 165 \\ e^{-\frac{(x-243.3)^2}{2(51.67)^2}}, & 165 \le x \le 300 \\ e^{-\frac{(x-366.2)^2}{2(51.67)^2}}, & 300 \le x \le 365 \end{cases} \tag{1}$$

**Table 1** Categorizing quadrants of months

| Quadrants | Months | | |
|---|---|---|---|
| 1st | January | February | March |
| 2nd | April | May | June |
| 3rd | July | August | September |
| 4th | October | November | December |

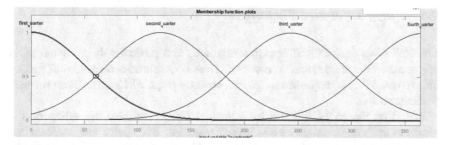

**Fig. 4** M1: Quadrant membership function

## 4.2 Category of Percentage of Infected People (PPI)

The PPI shown below is the function model of our membership function. Here each quadrant has been divided into three parts: low, medium and high in respect of year. To generate our membership function, we use triangular function in this model. The given parameters are observed from 2007 to 2015. The membership functions for PPI is given in Eq. 2 (Fig. 5).

Quadrant Membership Equation

$$
\mu_{PPI(\mathbf{X})} = \begin{cases}
0, & x \leq 0.000000804 \ (x \ is \ low) \\
\frac{0.000000804-x}{0.0000009796}, & 0.000000804 \leq x \leq 0.00000106 \ (x \ is \ low) \\
\frac{0.00000302-x}{0.00000196}, & 0.00000106 \leq x \leq 0.00000302 \ (x \ is \ low) \\
\frac{0.0000215-x}{0.0000042}, & 0.0000173 \leq x \leq 0.0000215 \ (x \ is \ medium) \\
\frac{0.0000794-x}{0.0000579}, & 0.0000264 \leq x \leq 0.000061 \ (x \ is \ high) \\
\frac{0.000061-x}{0.0000346}, & 0.0000215 \leq x \leq 0.0000794 \ (x \ is \ medium) \\
\frac{0.00013-x}{0.000069}, & 0.000061 \leq x \leq 0.00013 \ (x \ is \ high)
\end{cases} \qquad (2)
$$

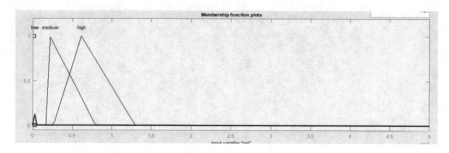

**Fig. 5** M2: PPI membership function

## 4.3   Fuzzy Inference Technique

Epidemic is an outgrown disease in a rural area at a particular time. Therefore, in this scenario the area should be constant. Time is Quadrant and chances of getting infected are PPI. From quadrants and PPI, we have acquired 12 antecedents for each particular result.

Based on the above idea we have made 10 rules and also depending on the results of epidemic chances we divide each result into three parts: low, medium and high. The rules are given in Table 2. Trapezoidal function has been used to justify low and high frequency and triangular function has been used for medium frequency of epidemic chances. The frequencies are the possibilities of dengue infection. The membership functions are given in Eq. 3. Note that we use Mamdani min and centroid function here to get our crisp values in the fuzzy algorithm (Fig. 6).

Epidemic Chance (EC) Membership Equation

$$
\mu_{EC(\mathbf{X})} = 
\begin{cases}
1, & 0 \leq x \leq 10 \ (x \ is \ Low) \\
\frac{50-x}{40}, & 10 \leq x \leq 40 \ (x \ is \ Low) \\
\frac{50-x}{20}, & 30 \leq x \leq 50 \ (x \ is \ Low) \\
\frac{70-x}{20}, & 50 \leq x \leq 70 \ (x \ is \ Medium) \\
\frac{99.96-x}{49.96}, & 50 \leq x \leq 99.96 \ (x \ is \ Medium) \\
1, & 90 \leq x \leq 100 \ (x \ is \ High)
\end{cases}
\tag{3}
$$

**Table 2** Fuzzy rules

|  | Percentage of People Infected (PPI) | | |
|---|---|---|---|
| Epidemic chance | $PPI_{Low}$ | $PPI_{Medium}$ | $PPI_{High}$ |
| First quadrant | low | low | low |
| Second quadrant | low | low | medium |
| Third quadrant | medium | high | high |
| Fourth quadrant | low | medium | medium |

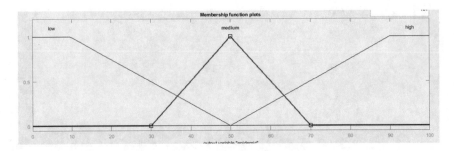

**Fig. 6** M3: Epidemic chance membership function

## 4.4 ANFIS

We have trained our dataset using ANFIS with of our 9 years' data of PPI along with four quadrants for each years from 2007 to 2015. Then we got our output values of PPI for any years along with quadrant. Using of values of PPI from our FIS we got our outcome results for epidemic chances. We have total 36 set of data for our ANFIS. Figures 7 and 8 depicts the training error and ANFIS structure.

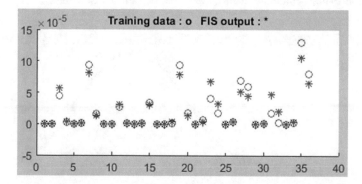

**Fig. 7** Trained dataset

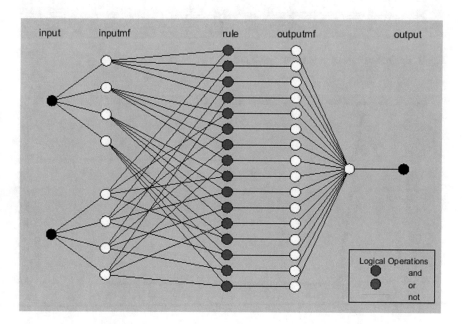

**Fig. 8** ANFIS structure

## 5  Result Analysis

We have found two different surface views from our FIS. One is quadrant versus epidemic another one is PPI versus epidemic. Figure 9 represents the epidemic chances in terms of season. Since we divide our days into four quadrants, it shows how days could affect epidemic chances.

Figure 10 represents the epidemic chances in terms of PPI (people per infected). It shows how increasing rate of infected people increase the epidemic chances.

Figure 11 represents the overall PPI_quadrants versus epidemic chances. It shows the overall epidemic chances in terms of PPI and quadrant.

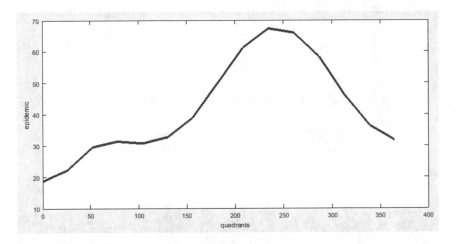

**Fig. 9**  Epidemic chances in terms of quadrants

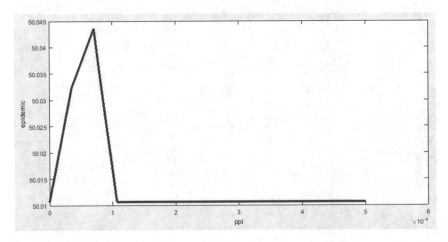

**Fig. 10**  Epidemic chances in terms of PPI

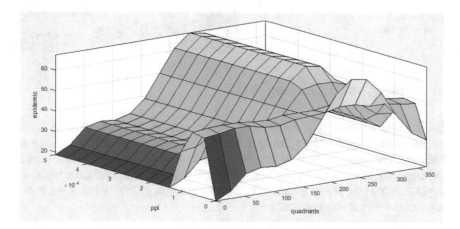

**Fig. 11** The overall PPI_quadrants versus epidemic chances

**Fig. 12** 24.1 % epidemic chances in 1995

**Fig. 13** 37.4 % epidemic chances in 1997

**Fig. 14** 49.7 % epidemic chances in 2001

**Fig. 15** 68.8 % epidemic chances in 2003

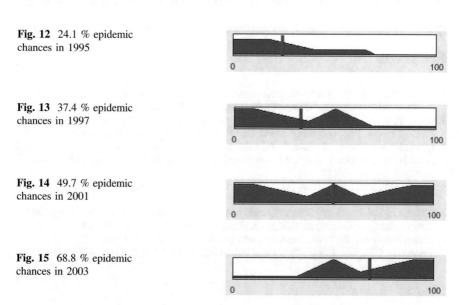

We will observe the epidemic chances from our FIS in terms of year 1995, 1997, 2001, 2003, 2005. In first quadrant of (Fig. 12) shows the epidemic chances 24.1 % which corresponds to the lower chance of epidemic.

Second quadrant of 1997 (Fig. 13) shows the epidemic chances of 37.4 % which also corresponds the lower chance.

Second quadrant of 2001 (Fig. 14) correspond 49.7 % epidemic chances.

Third quadrant of 2003 (Fig. 15) shows the epidemic chances of 68.8 %. which corresponds to the higher chance of dengue epidemic.

Finally, fourth quadrant of 2005 (Fig. 16) corresponds the lower chances of epidemic chances which is 31.6 %

**Fig. 16** 31.6 % epidemic
chances in 2005

## 6 Conclusion and Future Work

In this paper we have identified dengue epidemic chance in Dhaka only. In future
we can narrow our field to different thanas (sub-districts) to get the real picture of
epidemic situation. We have not considered different biological factors like water
logging, mosquito density etc. as inputs due to scarcity of data. With those inputs as
parameters we can surely improve our fuzzy algorithm so that we may get better
accuracy of identifying epidemic.

**Acknowledgment** The raw data set provided by IEDCR of monthly dengue cases helped us to
finish this paper. We retrieved population data of Dhaka from Bangladesh Bureau of Statistics.

## References

1. Chen, H.L., et al.: An efficient diagnosis system for detection of Parkinson's disease using fuzzy
   k-nearest neighbor approach. Expert Syst. Appl. **40**(1), 263–271 (2013)
2. Azmi, B.M.: Orchid disease detection using image processing and fuzzy logic IEEE Conf.
   Electr. Comput. Eng. (ICEESE), 37–42 (2013). doi:10.1109/ICEESE.2013.6895039
3. Longini, I.M., Yunus, M., Zaman, K., Siddique, A.K., Sack, R.B., Nizam, A.: Epidemic and
   endemic cholera trends over a 33-year period in Bangladesh. J. Infect. Dis. **186**, 246–251
   (2002). doi:10.1086/341206
4. Tuan, Q.D., Hiep, V.L., Tru, H.C., Quang, C.L., Hai, T.D.: Forecasting the magnitude of
   dengue in Southern Vietnam. ACIIDS (2016)
5. Institute of Epidemiology, Disease Control and Research Monthly Dengue Cases: http://www.
   iedcr.org/images/files/dengue/DEN_DKA_31.12.2015.pdf. Accessed 31 Dec 2015
6. Dhaka Population Research: http://www.bbs.gov.bd. Accessed 20 Feb 2016
7. Tsoukalas, L.H., Uhrig, R.E.: Fuzzy and Neural Approach in Engineering. Wiley, New York,
   NY (1997)

# MIMO Antenna Design for Future 5G Wireless Communication Systems

M. Aziz ul Haq, M. Arif Khan and Md Rafiqul Islam

**Abstract** This article presents a novel multiple input multiple output (MIMO) antenna design for future 5G wireless communication systems. An ultra-wideband antenna array is designed on a printed circuit board which is suitable for radio frequency circuits. Array consists of four antenna elements which are placed in a rectangular shape with separate excitation source. In multi antenna systems, separation between antennas is a challenging task which is achieved using corner truncation ground structure in the proposed design. This reduces mutual coupling between elements significantly. The measured impedance bandwidth of the proposed antenna array ranges from 3 to 9 GHz for a reflection coefficient of less than −10 dB. The proposed design also exhibits stable radiation patterns over the entire frequency band of interest.

## 1 Introduction

5G (Fifth Generation) is an emerging wireless technology that is expected to be commercialized by 2020. There has been a lot of interest in research and development of 5G wireless systems from both academia and commercial companies such as Huawei Technologies [1]. One of the major components of 5G technology is massive MIMO antenna system to achieve 10−100× times bandwidth as compared to current 4G and LTE-advanced wireless systems. Multiple antennas are used both at the transmitter and receiver to achieve promised goals in future wireless system. In this article, we present a simple but fundamental building block

M. Aziz ul Haq · M. Arif Khan (✉) · M.R. Islam
School of Computing and Mathematics, Charles Sturt University,
Wagga, NSW 2650, Australia
e-mail: mkhan@csu.edu.au

M. Aziz ul Haq
e-mail: muhammad.azizulhaq@yahoo.com

M.R. Islam
e-mail: mislam@csu.edu.au

© Springer International Publishing Switzerland 2016                 175
R. Lee (ed.), *Software Engineering, Artificial Intelligence, Networking and Parallel/Distributed Computing*, Studies in Computational Intelligence 653, DOI 10.1007/978-3-319-33810-1_14

of MIMO antenna systems that can be used as a basic building block for larger (massive) MIMO antenna arrays.

One of the major challenges with MIMO antennas is the mutual coupling between elements. Mutual coupling affects the performance of antennas and hence the performance of the overall system. In order to overcome mutual coupling problem, isolation between antennas should be high in the proposed design. Achieving high isolation becomes even difficult when designing small antennas, particularly for hand held devices with MIMO elements [1–6].

In most of the MIMO antenna literature, the antenna elements are made up of either printed monopoles or Planar Inverted F Antennas (PIFAs). It is difficult to achieve a reasonable isolation between antenna elements in such designs. Therefore, various isolation structures are used to achieve less isolation, but these structures make the design complex and even non-planar. In order to overcome such challenges, we present a printed triangular antenna (PTA) in this article which is operating at 6 GHz band to support massive MIMO operating frequency. The proposed antenna is simple in design, easy to fabricate and has a planar design as shown in Fig. 1. Furthermore, we do not use any separate isolation structure between antenna elements and achieve excellent isolation between antenna elements. The proposed design also provides good bandwidth compared to a patch antenna.

The rest of the paper is structured as follows. Section 2 presents the discussion on MIMO antenna design. Section 3 presents performance of the proposed MIMO antenna system. This section also presents both simulated and measured results. Finally, Sect. 4 concludes the paper.

**Fig. 1** Schematic layout of the proposed 4 elements MIMO antenna systems (*top view*)

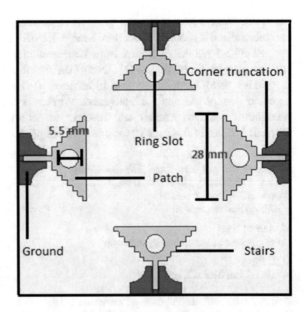

## 2 MIMO Antenna Design

Figure 1 shows the geometrical layout of the proposed MIMO antenna system. The design includes the radiator, ground plane, slots on the radiator structure and four microstrip transmission lines as shown in Fig. 1. The MIMO antenna system is designed using ANSYS HFSS [7] software as well as fabricated on FR-4 substrate and performance is measured. The photograph of the fabricated prototype MIMO antenna is shown above in Fig. 2.

The proposed MIMO antenna system is designed and simulated using commercially available electromagnetic simulation software HFSS. The designed antenna system has been fabricated for experimental validation of the simulated results, shown in Fig. 2. The MIMO antenna system is fabricated on FR-4 substrate with permittivity of $\varepsilon_r = 4.3$, loss tangent of $\delta = 0.025$, height, width and length dimensions of $1.6 \times 90 \times 90$ mm$^3$ respectively. The radiator and ground plan are on top and bottom of the dielectric substrate. Both are made of copper material with thickness of 0.035 mm and conductivity of $\sigma = 5.96 \times 10^7$ S/m. In order to achieve 50 $\Omega$ input impedance matching, a microstrip transmission line is used to excite the suitable set. The introduction of rectangular stairs on a radiator has made a great impact on the performance of antenna in terms of bandwidth and Return Loss (RL). The most important and interesting part in this design is the introduction of corner truncated ground plan of the MIMO antenna system. A slot cutting technique on radiating patch or ground is used to extend the frequency at high frequency to increase the bandwidth [8–11]. Overall, the proposed stair based feeder line technique with truncated ground structure is helpful to enhance the

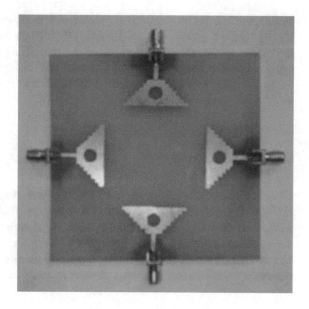

**Fig. 2** Fabricated prototype of the proposed MIMO antenna system

impedance matching of the proposed MIMO antenna and their effect on antenna performance is evident in Sect. 3 with various results.

## 2.1 Mutual Coupling

The basic requirement of reflection coefficient, $\Gamma$, of any antenna is less than $-10$ dB for practical use. In case of a MIMO system, as we have multiple antennas, hence there is a high probability of mutual coupling. This can be reduced by using some innovative techniques in antenna design. There are two commonly used techniques in this regard. The first technique takes advantage of the perpendicular structure of the ports; since the simplest method to decrease coupling between adjacent antennas is to move them farther apart from each other and orient them in such a way that their polarization is mismatched. The second technique is the corner truncation at the ground plane. This technique is commonly used to extend the bandwidth of the antenna. In this paper, we used this second technique to achieve higher bandwidth of the proposed MIMO antenna system. We tested the truncation for different sizes of circles ranging from 4 to 6 mm radius. The best result was observed with the truncation of 5.5 mm radius circle. The change in the antenna performance is due to the change in current distribution of the proposed antenna.

## 3 Characterization of the Proposed MIMO Antenna System

In this section, we present the simulated and measured results of the proposed MIMO antenna system. We used HFSS [7] software to simulate the antenna and then tested measured results in an echoic chamber. The parameters of interest in this study are the antenna S-Parameters, Gain and radiation patterns. We present a study on each parameter for the proposed MIMO antenna system separately in the following sub-sections.

## 3.1 S-Parameters

The measured results of the proposed MIMO antenna system are taken on Agilent PNA-X N5242A network analyzer. A comparison all simulated and measured $S_{ij}$ parameters are shown in Figs. 3, 4, 5 and 6. $S$ parameters are important in antenna design since they show power transmitted through and reflected back from each antenna with respect to all other antennas. It is clear from the simulated and measured results that each antenna element resonates with the bandwidth of 6 GHz

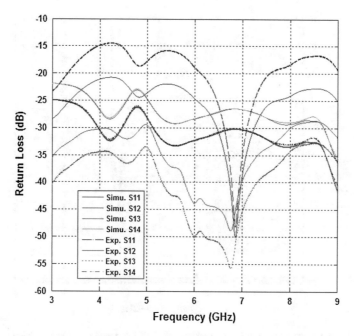

**Fig. 3** Simulated and measured S-parameters of the 1st antenna element with respect to 2nd, 3rd and 4th antenna elements in the proposed MIMO system

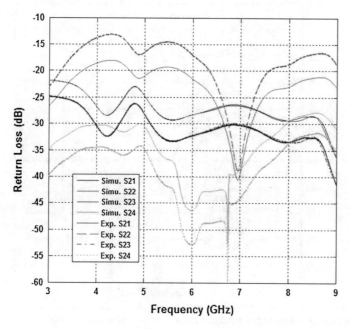

**Fig. 4** Simulated and measured S-parameters of the 2nd antenna element with respect to 1st, 3rd and 4th antenna elements in the proposed MIMO system

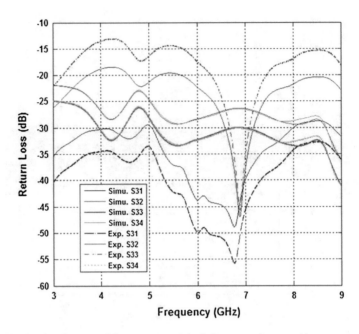

**Fig. 5** Simulated and measured S-parameters of the 3rd antenna element with respect to 1st, 2nd and 4th antenna elements in the proposed MIMO system

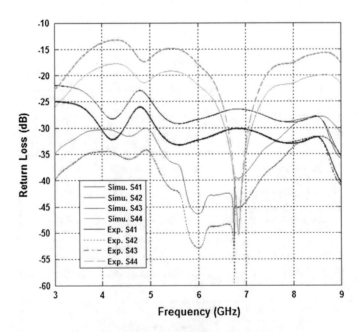

**Fig. 6** Simulated and measured S-parameters of the 4th antenna element with respect to 1st, 2nd and 3rd antenna elements in the proposed MIMO system

**Table 1** Peak gain ($G_p$) and radiation efficiency ($\eta$) of the proposed MIMO antenna system

| Element # | Peak gain ($G_p$) (dB) | Radiation efficiency ($\eta$) (%) |
|-----------|------------------------|-----------------------------------|
| 1 | 11.50 | 98.0 |
| 2 | 11.00 | 97.8 |
| 3 | 12.00 | 98.0 |
| 4 | 11.09 | 97.0 |

covering the 5G wireless communication system band. It is also evident that measured results exhibit a very good agreement with the simulations.

## 3.2 Gain of the Proposed MIMO Antenna System

The peak gain, $G_p$, and radiation efficiency, $\eta$, of each MIMO antenna element in the system are obtained by exciting the specific element and terminating other elements with 50 $\Omega$ matched loads. The peak gain and radiation efficiency of each antenna element are summarized in Table 1. It can be seen that all antenna elements have peak gains varying from 11 to 12.5 dB with gain variations of less than 1 dB. The calculated radiation efficiencies of the four antenna elements are varying from 97 to 98 % with 1 % variation at 7 GHz.

## 3.3 Radiation Patterns

Polarization diversity is a technique where different polarizations (vertical or horizontal) are used to improve the performance of MIMO systems. According to the design structure, antenna ports are orthogonal to each other, therefore horizontal and vertical polarizations were obtained. Measured co- and cross-polarization 2-D radiation patterns of the proposed MIMO antenna system at 6 GHz are shown in Fig. 7. The normalized gains are plotted in xz-plane and yz-plane for ports 1 and 2. The data was collected in anechoic chamber with one port excited and the other ports terminated with a standard 50 $\Omega$ matched load. According to the symmetric structure of the proposed MIMO antenna system, the radiation patterns of ports 3 and 4 are not shown as they are the same as ports 1 and 2 respectively. Since the ports are orthogonal, a dual polarized antenna has been achieved. Therefore, the vertical and horizontal electric fields are produced from ports 1 and 2, respectively. Moreover, an omnidirectional radiation pattern without any deep nulls is obtained at the expected operating frequencies.

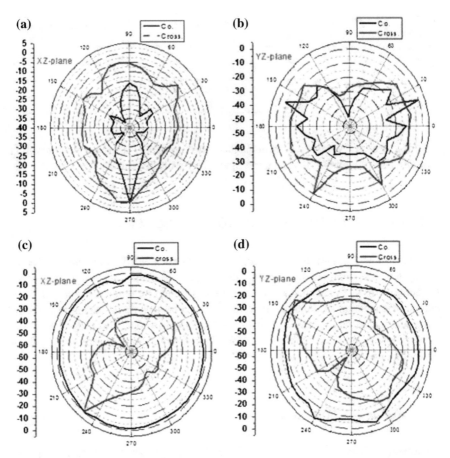

**Fig. 7** Measured co- and cross-polarization of radiation patterns of the proposed MIMO antenna system at 7 GHz. **a** xz-plane port 1, **b** yz-plane port 1, **c** xz-plane port 2 and **d** yz-plane port 2

## 4 Conclusions

In this paper, a four elements MIMO antenna system is designed, simulated and presented on FR-4 substrate for future 5G wireless system. The electromagnetic properties of the proposed MIMO system are tested and measured results of the fabricated antenna are used to confirm the simulated results. The measured impedance bandwidth of the proposed MIMO antenna array ranges from 3 to 9 GHz for a reflection coefficient of less than −10 dB which covers the operating frequency of 5G. Furthermore, the proposed design exhibits stable radiation patterns over the whole frequency band of interest. Due to very good performance of the proposed antenna system and its miniaturize size, it can be considered as a good candidate for 5G communication applications, particularly for hand held devices.

# References

1. Tong, W., Peiying, Z.: 5G: A technology vision. Huawei Technical Publications (2016). http://www1.huawei.com/en/about-huawei/publications/winwin-magazine/hw-329304.htm
2. Gupta, A., Jha, R.K.: A survey of 5G network: architecture and emerging technologies. IEEE Access **3**, 1206–1232 (2015)
3. Zhao, L., Yeung, L.K., Ke-Li, Wu: A coupled resonator decoupling network for two-element compact antenna arrays in mobile terminals. IEEE Trans. Antennas Propag. (2014). doi:10.1109/TAP.2014.2308547
4. Cihangir, A., Ferrero, F., Jacquemod, G., et al.: Neutralized coupling elements for MIMO operation in 4G mobile terminals. IEEE Antennas Wirel. Propag. Lett. (2014). doi:10.1109/LAWP.2014.2298392
5. Al-Hadi, A., Tian, R.: Impact of multiantenna real estate on diversity and MIMO performance in mobile terminals. IEEE Antennas Wirel. Propag. Lett. (2013). doi:10.1109/LAWP.2013.2293596
6. Zhang, S., Lau, B.K., Tan, Y., et al.: Mutual coupling reduction of two PIFAs with a T-shape slot impedance transformer for MIMO mobile terminals. IEEE Trans. Antennas Propag. (2012). doi:10.1109/TAP.2011.2180329
7. http://www.ansys.com/Products/Electronics/ANSYS-HFSS
8. Sharawi, M.S., Numan, A.B., Khan, M.U., et al.: A dual element dual-band MIMO antenna system with enhanced isolation for mobile terminals. IEEE Antennas Wirel. Propag. Lett. (2012). doi:10.1109/LAWP.2012.2214433
9. Jiang-Yi, P., Shao-Qiu, X., Zhuo-Fu, D., et al.: Two element PIFA antenna system with inherent performance of low mutual coupling. IEEE Antennas Wirel. Propag. Lett. (2009). doi:10.1109/LAWP.2009.2035454
10. Sarkar, P., Chowdhury, S.K.: A novel compact, microstrip antenna with multi-frequency operation. In: International Seminar/Workshop on Direct and Inverse Problems of Electromagnetic and Acoustic Wave Theory (DIPED) (2009)
11. Polvka, M.: Multiband behavior of the rectangular microstrip patch antenna modied by T-notch perturbation elements. In: Proceedings of 18th International Conference on Applied Electromagnetics and Communications (ICECom) (2005)
12. Salonen, P., Keskilammi, M., Kivikoski, M.: New slot congurations for dual-band planar inverted F antenna. Microwave Opt. Technol. Lett. **28**(5), 293–298 (2001)
13. Haq, M.A., Khan, M.A.: A multiple ring slots ultra wide band antenna (MRS-UWB) for biomedical applications. In: 17th IEEE International Multi-Topic Conference (INMIC), pp. 56–60 (2014)

# Author Index

© Springer International Publishing Switzerland 2016                                          185
R. Lee (ed.), *Software Engineering, Artificial Intelligence, Networking
and Parallel/Distributed Computing*, Studies in Computational
Intelligence 653, DOI 10.1007/978-3-319-33810-1

Printed in the United States
By Bookmasters